Springer Japan KK

M. Omine, T. Kinoshita (Eds.)

Paroxysmal Nocturnal Hemoglobinuria and Related Disorders

Molecular Aspects of Pathogenesis

With 46 Figures

Springer

ISBN 978-4-431-68004-8 ISBN 978-4-431-67867-0 (eBook)
DOI 10.1007/978-4-431-67867-0

Library of Congress Cataloging-in-Publication Data applied for.

Printed on acid-free paper

Originally published by Springer-Verlag Tokyo in 2003
Softcover reprint of the hardcover 1st edition 2003

Typesetting: Camera-ready by the editors and authors

SPIN: 10855431

Foreword

Paroxysmal nocturnal hemoglobinuria (PNH) has been designated an intractable disease by the Japanese Government since 1974. In the succeeding years, the Research Committee for Idiopathic Hematopoietic Disorders has, with considerable success, concentrated on accumulating clinical data and elucidating the biological and biochemical pathogenesis of PNH.

Since its establishment in 1973, the Japan Intractable Diseases Research Foundation has supported research on intractable diseases in Japan. This year our Program Committee decided to support an International Symposium on PNH and related disorders. One reason we chose this disease as the main subject of the Symposium was that we consider PNH to be one of the most important intractable diseases, and another reason is that Professor T. Kinoshita and his group recently discovered a novel genetic change in PNH patients. This discovery marked the beginning of an epoch and opened a new horizon to clinical as well as experimental investigations, which may be expected to bring about better management of the disease.

Consequently, we asked Professor M. Omine, Chairman of the Research Committee, to organize the Symposium in collaboration with Professor T. Kinoshita. The time and venue were planned so that the Symposium could be held in close connection with the 30th Annual Scientific Meeting of the International Society for Experimental Hematology, to facilitate the participation of foreign scientists.

It is my strong belief that this Symposium will be scientifically fruitful, and as Chairman of the Program Committee of the Foundation I should like to thank all the participants for making this Symposium a memorable one.

Fumimaro Takaku, M.D.
Acting President
Japan Intractable Diseases
Research Foundation
August 2001

Preface

To hematologists, paroxysmal nocturnal hemoglobinuria (PNH) has been a mystery disease for more than a century. Slow and steady, sometimes painstaking accumulation of scientific knowledge on the unique nature of PNH blood cells culminated in the discovery of the *PIG-A* gene mutation by T. Kinoshita and his associates in 1993. Since then, many important scientific observations have followed at an amazing tempo in many frontier laboratories all over the world, resulting in extensive confirmation and bringing about a new paradigm regarding PNH and related disorders. By the close of the last century, a new concept of PNH had prevailed, and clinical characteristics had been reinterpreted in detail on the basis of renewed molecular biology. At the same time, new discoveries have created many new questions. Among others, the mechanism of clonal expansion of PNH cells, the emergence of abnormal clones under bone marrow failure such as aplastic anemia and myelodysplastic syndromes, and thrombotic complications largely remain unclarified. As for treatment, major problems are still left unsolved. As an aggressive treatment, hematopoietic stem cell transplantation and immunosuppresive therapy have been chosen under certain circumstances, but treatment results in general are far from satisfactory. New approaches, including gene therapy and application of complement inhibitors, are eagerly awaited.

We thought now would be most timely and significant for interested scientists to get together in a one-and-a-half-day symposium, to communicate with one another and to summarize the current understanding of the molecular pathogenesis of PNH and related disorders. This gathering would be meaningful for constructing better strategies through international collaboration for further clarification of the disease and for development of new treatment approaches. Upon approval by the Japan Intractable Diseases Research Foundation, the Program Committee elaborated a framework for the Symposium and decided to invite 10 distinguished experts from abroad and 7 Japanese investigators for lecture presentations. All invitations were willingly accepted. In addition, 14 papers were submitted upon request as free communications. The Organizing Committee wish to thank all the participants for their enthusiasm and efforts dedicated to this Symposium.

This monograph thus represents the proceedings and the free communications of the Symposium. The reader will find this book most useful in obtaining a knowledge of up-to-date concepts and experts' opinions in this rapidly evolving area of hematology.

Mitsuhiro Omine, M.D.
Chairman of the Organizing Committee
Taroh Kinoshita, Ph.D.
Chairman of the Program Committee
Summer 2001

International Symposium on PNH and Related Disorders: Molecular Aspects of Pathogenesis

Keio Plaza Inter-Continental Tokyo, August 28–29, 2001

Sponsored by the Japan Intractable Diseases Research Foundation

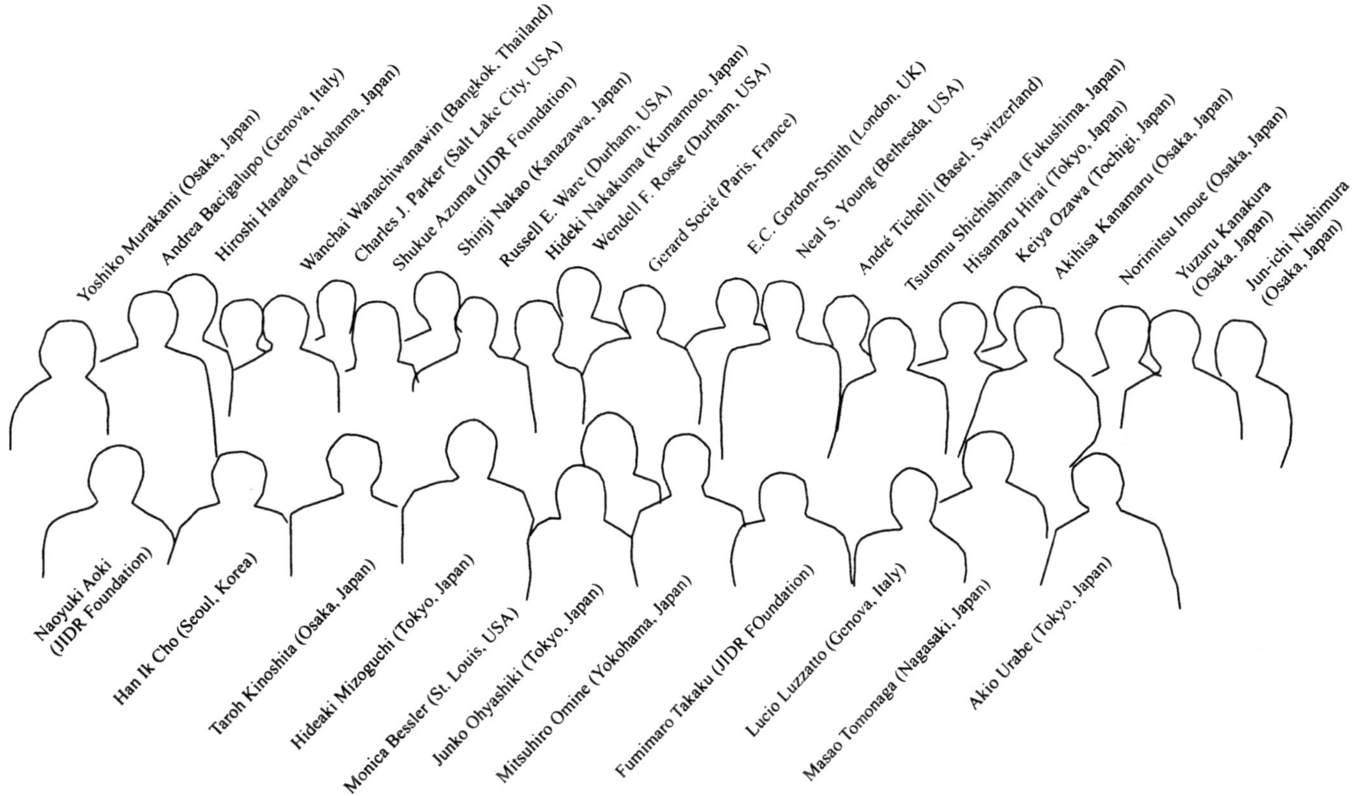
Yoshiko Murakami (Osaka, Japan)
Andrea Bacigalupo (Genova, Italy)
Hiroshi Harada (Yokohama, Japan)
Wanchai Wanachiwanawin (Bangkok, Thailand)
Charles J. Parker (Salt Lake City, USA)
Shukue Azuma (JIDR Foundation)
Shinji Nakao (Kanazawa, Japan)
Russell E. Ware (Durham, USA)
Hideki Nakakuma (Kumamoto, Japan)
Wendell F. Rosse (Durham, USA)
Gerard Socié (Paris, France)
E.C. Gordon-Smith (London, UK)
Neal S. Young (Bethesda, USA)
André Tichelli (Basel, Switzerland)
Tsutomu Shichishima (Fukushima, Japan)
Hisamaru Hirai (Tokyo, Japan)
Keiya Ozawa (Tochigi, Japan)
Akihisa Kanamaru (Osaka, Japan)
Norimitsu Inoue (Osaka, Japan)
Yuzuru Kanakura
(Osaka, Japan)
Jun-ichi Nishimura
(Osaka, Japan)
Naoyuki Aoki
(JIDR Foundation)
Han Ik Cho (Seoul, Korea)
Taroh Kinoshita (Osaka, Japan)
Hideaki Mizoguchi (Tokyo, Japan)
Monica Bessler (St. Louis, USA)
Junko Ohyashiki (Tokyo, Japan)
Mitsuhiro Omine (Yokohama, Japan)
Fumimaro Takaku (JIDR FOundation)
Lucio Luzzatto (Genova, Italy)
Masao Tomonaga (Nagasaki, Japan)
Akio Urabe (Tokyo, Japan)

Organization of Symposium

ACTING PRESIDENT, JAPAN INTRACTABLE DISEASES RESEARCH FOUNDATION

Fumimaro Takaku	Jichi Medical School, Tochigi

CHAIRMAN

Mitsuhiro Omine	Division of Hematology, Showa University Fujigaoka Hospital, Yokohama

PROGRAM COMMITTEE CHAIRMAN

Taroh Kinoshita	Department of Immunoregulation Research Institute for Microbial Diseases Osaka University, Osaka

ORGANIZING COMMITTEE

Hisamaru Hirai	Department of Hematology and Oncology University of Tokyo, Tokyo
Yuzuru Kanakura	Department of Hematology and Oncology Osaka University, Osaka
Akihisa Kanamaru	Third Department of Medicine Kinki University, Osaka
Hideaki Mizoguchi	Department of Hematology, Tokyo Women's Medical University, Tokyo
Shinji Nakao	Third Department of Internal Medicine Kanazawa University, Kanazawa
Keiya Ozawa	Division of Hematology, Department of Medicine, Jichi Medical School, Tochigi
Masao Tomonaga	Department of Hematology Atomic Bomb Disease Institute Nagasaki University, Nagasaki
Takashi Uchiyama	Department of Hematology/Oncology Kyoto University, Kyoto
Akio Urabe	Division of Hematology NTT Kanto Medical Center, Tokyo

SECRETARY-GENERAL

Shukue Azuma	Japan Intractable Diseases Research Foundation

Acknowledgments

The chairman and members of the Organizing Committee wish to thank the following organizations for their generous support of this Symposium:

The Japan Intractable Diseases Research Foundation

Research Committee for Idiopathic Hematopoietic Disorders
The Ministry of Health, Labour and Welfare, Japan

Japanese Society of Hematology

Japanese Society of Clinical Hematology

Contents

Part 3: Clinical Pathology and Natural History of PNH

Part 4: Mechanism of Clonal Expansion in PNH

Part 5: PNH Clones Under Bone Marrow Failure

Part 6: PNH, Aplastic Anemia, Myelodysplastic Syndromes (1): Pathology

Part 7: PNH, Aplastic Anemia, Myelodysplastic Syndromes (2): Clinical Aspects

Part 8: Related Topics

Contributors

Part 1

Overview of Paroxysmal Nocturnal Hemoglobinuria

Overview of Paroxysmal Nocturnal Hemoglobinuria: Molecular Genetics

Taroh Kinoshita

Research Institute for Microbial Diseases, Osaka University, 3-1 Yamada-oka, Suita, Osaka 565-0871, Japan

Summary: Paroxysmal nocturnal hemoglobinuria (PNH) is characterized by clonal blood cells that are defective in the surface expression of GPI-anchored proteins. Glycosylphosphatidylinositol (GPI) is synthesized in the endoplasmic reticulum and is transferred en bloc to the protein. Defective GPI-anchor biosynthesis results in a lack of the cell surface expression of GPI-anchored proteins. Affected cells from patients with PNH are defective in the first reaction of GPI biosynthesis. This defect is due to somatic mutation in PIG-A gene that encodes a subunit of the N-acetylglucosamine transferase. PIG-A is localized on the X-chromosome. Therefore, one inactivating mutation causes the defective biosynthesis of GPI.

More than twenty genes are involved in biosynthesis of GPI. In more than 150 patients with PNH characterized at the molecular level, PIG-A is responsible for GPI-anchor deficiency. This should be due to the X-linkage of PIG-A gene and autosomal localization of all other genes involved in biosynthesis of GPI.

Somatic mutation of PIG-A occurs in the hematopoietic stem cell. The same clones have been maintained for at least six years. In many patients, more than one mutant clones are found. Usually, one is predominant and others are minor. Somatic mutation of PIG-A is found in a very small fraction of granulocytes from most of the healthy individuals and in a bigger fraction of granulocytes from nearly 90% of the untreated patients with aplastic anemia. More than 10% of patients with aplastic anemia have clearly recognizable PIG-A mutant clone. These findings suggest that somatic mutation of PIG-A occurs in normal individuals and that clonal expansion requires extra event(s), such as selection by some pathological condition and/or some additional genetic change.

Key words: GPI-anchor, hematopoietic stem cell, somatic mutation, X-chromosome, aplastic anemia

Introduction

PNH is characterized by clonal blood cells that are defective in the surface expression of GPI-anchored proteins, such as CD59 and decay accelerating factor (DAF or CD55). Clonal cells defective in GPI-anchored proteins appear in all blood cell types but not in other tissues, indicating that a clone of the multipotential hematopoietic stem cell defective in GPI-anchored proteins is present in patients with PNH. The first step of pathogenesis of PNH would, therefore, be the generation of such a clone by somatic mutation. The second step of pathogenesis is expansion of the clone to the size big enough for clinical manifestation. Third, disease persists for many years, often for more than 20 years.

CD59 and DAF are complement regulatory proteins that protect host cells from the destructive action of complement. Red cells are protected mainly with CD59 and DAF, so that affected red cells of patients with PNH are abnormally sensitive to complement. Upon activation of complement during infections and other events, affected red cells are lysed, resulting in hemolytic anemia and hemoglobinuria [1].

Biosynthesis of GPI-anchored proteins and its deficiency in PNH

The biochemical basis of PNH abnormality is a defect in biosynthesis of GPI, a complex glycolipid that is covalently attached to the C-terminus of a wide variety of proteins, thereby, acts as a membrane-binding portion of those proteins. GPI is synthesized in the endoplasmic reticulum through nine or more reactions and is transferred en bloc to the protein by GPI transamidase [2]. If GPI-anchor is not attached, the proteins are either degraded or secreted into the extracellular space, resulting in a lack of the cell surface expression of GPI-anchored proteins. In fact, at least 19 different proteins that are normally GPI-anchored are deficient on various blood cells from patients with PNH [1].

Studies with affected cell lines from patients with PNH demonstrated a defect in the first reaction of GPI biosynthesis, a transfer of N-acetylglucosamine from uridinediphosphate-N-acetylglucosamine to phosphatidylinositol [3-6]. An N-acetylglucosamine transferase responsible for this reaction consists of at least five proteins, PIG-A, PIG-C, PIG-H, PIG-P and GPI1 [7, 8]. The affected cells are defective in the gene that encodes PIG-A [9], a catalytic subunit of the enzyme [10].

The defect is caused by somatic mutation in PIG-A gene [9]. PIG-A is localized on the X-chromosome at Xp22.1 [9, 11]. Therefore, both male and female somatic cells have only one functional copy of PIG-A gene. One inactivating somatic mutation causes a loss of PIG-A function in male cells as well as in female cells, resulting in the defective biosynthesis of GPI.

More than twenty genes are involved in biosynthesis of GPI and its transfer to proteins. To date, nineteen genes have been cloned in human system [2]. PIG-A is the only X-linked gene among the nineteen genes (Table 1). Remarkably, in more than 150 patients with PNH characterized at the molecular level, PIG-A is responsible for GPI-anchor deficiency [12, 13]. This should be due to the X-linkage of PIG-A gene and autosomal localization of all other genes involved in biosynthesis and transfer of GPI.

Somatic mutation of PIG-A gene

The somatic mutations are mainly small mutations [12, 13]. They are various among patients and distributed widely in coding regions and splice sites. Single base substitutions and single base deletions contribute one-third each. The last one-third consists of single base insertions, two to several bases insertions and deletions, deletion/insertions involving several bases, and a few cases of larger deletions and insertions [12, 13].

Majority of the somatic mutations cause frame-shift, resulting in premature stop codons. Base substitutions are non-sense mutations and missense mutations. Some of the missense muattions result in a partial loss of PIG-A activity that accounts for a partial deficiency of GPI-anchored proteins [12, 13].

Somatic mutation of PIG-A occurs in the multipotential hematopoietic stem cell. Affected granulocytes and lymphocytes from individual patients have the same mutation [9].

The same clones have been maintained for at least six years (see the chapter by T. Hirota, this volume). In many patients, more than one mutant clones are found [14, 15]. Usually, one is predominant and others are minor.

Clonal expansion of PIG-A mutant cells

Somatic mutation of PIG-A is found in a very small fraction (average 0.002%) of granulocytes from most of the healthy individuals [16]. Whether these mutations occurred in the hematopoietic stem cells is unclear. PIG-A mutation is also found in a bigger fraction (>0.003%) of granulocytes from nearly 90% of the untreated patients with aplastic anemia [17]. More than 10% of patients with aplastic anemia have clearly recognizable PIG-A mutant clone [18-20]. PIG-A mutant clones are also found in patients with myelodysplasia [21, 22]. These findings together suggest that somatic mutation of PIG-A occurs in normal individuals and that clonal expansion requires extra event(s).

The notion that PIG-A mutant clone does not expand autonomously is supported by data obtained from mouse models. In mice having GPI-deficient hematopoietic stem cells generated by disruption of Pig-a gene (mouse orthologue

Table 1. Chromosomal location of genes involved in biosynthesis and transfer of GPI

No	Gene	Location
1	PIG-A	Xp22.1
2	PIG-C	1q23.3
3	PIG-H	14q11-q24
4	PIG-P	21q22.2
5	GPI1 (PIGQ)	16p13.3
6	PIG-L	17p12
7	PIG-M	1q22
8	PIG-N	18q21
9	PIG-B	15q21-q22
10	PIG-F	2p16-p21
11	PIG-O	9
12	GPI8 (PIGK)	1p22.2-p22.3
13	GAA1 (GPAA1)	8q24.3
14	PIG-S	17
15	PIG-T	20q12-q13
16	DPM1	20q13.1
17	DPM2	9q33
18	DPM3	1q21.2
19	SL15 (MPDU1)	17p13.1

of PIG-A), GPI-deficient cells did not expand [23-27] (see the chapter by M. Bessler, this volume, for further discussion).

Regarding the mechanism of clonal expansion of PIG-A mutant stem cells, there are two major hypotheses. First is selection by some pathological conditions [28-32]. This hypothesis is consistent with a highly frequent association between PNH and aplastic anemia. The major cause of aplastic anemia is very likely antoimmunity against hematopoietic stem cells. It is conceivable that if the effector of autoimmunity is an antibody that recognizes some GPI-anchored protein, GPI-anchor-deficient cells would escape and survive the autoimmune attack, resulting in clonal expansion. Or, if the effector is cytotoxic cells, GPI-anchor-deficient cells may be more resistant to cell-mediated cytotoxicity than normal cells. This could be either due to a lack of GPI-anchored proteins involved in cell-to-cell interactions or to inefficient presentation of target antigen(s) (see the chapters by L. Luzzatto et al, H. Nakakuma et al and Y. Murakami et al, in this volume).

The higher percentage of GPI-anchor-deficient granulocytes in patients with aplastic anemia may be simply accounted for by the decreased number of hematopoietic stem cells. However, when the percentage of GPI-anchor-deficient granulocytes exceeds several percent, it might be a result of selection.

The second hypothesis is that PIG-A mutant stem cell has other genetic change(s). Expanded PIG-A mutant cell clone can be regarded as a benign tumor (see the chapter by N. Inoue, this volume). It is possible that some additional genetic change(s) confer PIG-A mutant stem cells with a growth phenotype. This hypothesis is in line with the multi-step model of tumorigenesis. Occurrence of acute myelogenous leukemia in PIG-A mutant cells seen in patients with PNH [1]can be understood in such context.

Two hypotheses are not mutually exclusive. Rather, it seems possible that both mechanisms operate together. If PIG-A mutant stem cells survive immunological attacks due to a lack of GPI-anchor, they will proliferate much more frequently than usual to compensate anemic conditions. Elevated proliferation would increase a chance of further mutation and a generation of a subclone bearing growth phenotype. Immunological and molecular mechanisms of clonal expansion are under active investigations.

References

1. Rosse W F (2001) New insights into paroxysmal nocturnal hemoglobinuria. Cur Opi Hematol 8:61-67
2. Kinoshita T, Inoue N (2000) Dissecting and manipulating the pathway for glycosylphosphatidylinositol-anchor biosynthesis. Cur Opi Chem Biol 4:632-638
3. Takahashi M, Takeda J, Hirose S, et al. (1993) Deficient biosynthesis of N-acetylglucosaminyl phosphatidylinositol, the first intermediate of glycosyl phosphatidylinositol anchor biosynthesis, in cell lines established from patients with paroxysmal nocturnal hemoglobinuria J Exp Med 177:517-521
4. Hillmen P, Bessler M, Mason P J, et al. (1993) Specific defect in N-acetylglucosamine incorporation in the biosynthesis of the glycosylphosphatidylinositol anchor in cloned cell lines from patients with paroxysmal nocturnal hemoglobinuria Proc Natl Acad Sci USA 90:5272-5276
5. Armstrong C, Schubert J, Ueda E, et al. (1992) Affected paroxysmal nocturnal hemoglobinuria T lymphocytes harbor a common defect in assembly of N-acetyl-D-glucosamine inositol phospholipid corresponding to that in class A Thy-1 murine lymphoma mutants J Biol Chem 267:25347-25351
6. Hidaka M, Nagakura S, Horikawa K, et al. (1993) Impaired glycosylation of glycosylphosphatidylinositol-anchor synthesis in paroxysmal nocturnal hemoglobinuria leukocytes Biochem Biophys Res Commun 191:571-579
7. Watanabe R, Inoue N, Westfall B, et al. (1998) The first step of glycosylphosphatidylinositol biosynthesis is mediated by a complex of PIG-A, PIG-H, PIG-C and GPI1. EMBO J 17:877-885
8. Watanabe R, Murakami Y, Marmor M D, et al. (2000) Initial enzyme for glycosylphosphatidylinositol biosynthesis requires PIG-P and is regulated by DPM2. EMBO J 19:4402-4411

9. Takeda J, Miyata T, Kawagoe K, et al. (1993) Deficiency of the GPI anchor caused by a somatic mutation of the PIG-A gene in paroxysmal nocturnal hemoglobinuria Cell 73:703-711
10. Kostova Z, Rancour D M, Menon A K, et al. (2000) Photoaffinity labelling with P^3-(4-azidoanilido)uridine 5'-triphosphate identifies Gpi3p as the UDP-GlcNAc-binding subunit of the enzyme that catalyses formation of GlcNAc-phosphatidylinositol, the first glycolipid intermediate in glycosylphosphatidylinositol synthesis. Biochem J 350:815-822
11. Bessler M, Hillmen P, Longo L, et al. (1994) Genomic organization of the X-linked gene (PIG-A) that is mutated in paroxysmal nocturnal haemoglobinuria and of a related autosomal pseudogene mapped to 12q21. Hum Mol Genet 3:751-757
12. Kinoshita T, Inoue N, Takeda J (1995) Defective glycosyl phosphatidylinositol anchor synthesis and paroxysmal nocturnal hemoglobinuria Adv. Immunol. 60:57-103
13. Luzzatto L, Nafa K (2000) Genetics of PNH. In: N. S. Young and J. Moss (eds) Paroxysmal nocturnal hemoglobinuria and the glycosylphosphatidylinositol-linked proteins. Academic Press, San Diego, pp21-47
14. Endo M, Ware R, Vreeke T, et al. (1996) Molecular basis of the heterogeneity of expression of glycosyl phosphatidylinositol anchored proteins in paroxysmal nocturnal hemoglobinuria. Blood 87:2546-2557
15. Nishimura J, Inoue N, Wada H, et al. (1997) A patient with paroxysmal nocturnal hemoglobinuria bearing four independent PIG-A mutant clones. Blood 89:3470-3476
16. Araten D J, Nafa K, Pakdeesuwan K, et al. (1999) Clonal populations of hematopoietic cells with paroxysmal nocturnal hemoglobinuria genotype and phenotype are present in normal individuals. Proc Natl Acad Sci USA 96:5209-5214
17. Wang H, Chuhjo T, Yamazaki H, et al. (2001) Relative increase of granulocytes with a paroxysmal nocturnal haemoglobinuria phenotype in aplastic anaemia patients: the high prevalence at diagnosis. Eur. J. Haematol. 66:200-205
18. Nagarajan S, Brodsky R A, Young N S, et al. (1995) Genetic defects underlying paroxysmal nocturnal hemoglobinuria that arises out of aplastic anemia. Blood 86:4656-4661
19. Azenishi Y, Ueda E, Machii T, et al. (1999) CD59-deficient blood cells and PIG-A gene abnormalities in Japanese patients with aplastic anaemia. Br J Haematol 104:523-529
20. Schrezenmeier H, Hildebrand A, Rojewski M, et al. (2000) Paroxysmal nocturnal haemoglobinuria: a replacement of haematopoietic tissue? Acta Haematol 103:41-48
21. Iwanaga M, Furukawa K, Amenomori T, et al. (1998) Paroxysmal nocturnal haemoglobinuria clones in patients with myelodysplastic syndromes. Br J Haematol 102:465-474
22. Dunn D E, Tanawattanacharoen P, Boccuni P, et al. (1999) Paroxysmal nocturnal hemoglobinuria cells in patients with bone marrow failure syndromes. Ann Intern Med 131:401-408
23. Kawagoe K, Kitamura D, Okabe M, et al. (1996) GPI-anchor deficient mice: Implications for clonal dominance of mutant cells in paroxysmal nocturnal hemoglobinuria. Blood 87:3600-3606

24. Rosti V, Tremmi G, Soares V, et al. (1997) Murine embryonic stem cells without pig-a gene activity are competent for hematopoiesis with the PNH phenotype but not for clonal expansion. J. Clin. Invest. 100:1028-1036
25. Tremml G, Dominguez C, Rosti V, et al. (1999) Increased sensitivity to complement and a decreased red cell life span in mice mosaic for a non-functional Piga gene. Blood 94:2945-2954
26. Keller P, Tremml G, Rosti V, et al. (1999) X inactivation and somatic cell selection rescue female mice carrying a Piga-null mutation. Proc Natl Acad Sci USA 96:7479-7483
27. Keller P, Payne J L, Tremml G, et al. (2001) FES-Cre targets phosphatidylinositol glycan class A (PIGA) inactivation to hamatopoietic stem cells in the bone marrow. J Exp Med 194:581-589
28. Rotoli B, Luzzatto L (1989) Paroxysmal nocturnal haemoglobinuria Baillieres Clin Haematol 2:113-138
29. Young N S (1992) The problem of clonality in aplastic anemia: Dr Dameshek's riddle, restated. Blood 79:1385-1392
30. Luzzatto L, Bessler M, Rotoli B (1997) Somatic mutations in paroxysmal nocturnal hemoglobinuria: A blessing in disguise? Cell 88:1-4
31. Karadimitris A, Luzzatto L (2001) The cellular pathogenesis of paroxysmal nocturnal haemoglobinuria. Leukemia 15:1148-1152
32. Young N S (2000) Hematopoietic cell destruction by immune mechanisms in acquired aplastic anemia. Semin Hematol 37:3-14

An Overview of the Development of Specific Inhibitors of Complement: Opportunities for Therapy of Paroxysmal Nocturnal Hemoglobinuria

Charles J. Parker, M. D.

Hematology/Oncology Section VA Medical Center and The University of Utah School of Medicine Salt Lake City, Utah

Abstract: The erythrocytes of PNH are abnormally sensitive to complement-mediated lysis because they lack the two proteins (CD55 and CD59) that normally control the activity of complement on the cell surface. The hemolytic anemia of PNH is complement-mediated, but the mechanism by which complement induces spontaneous lysis of PNH RBC in vivo is speculative. Further, the basis of the greater hemolysis that occurs during sleep remains enigmatic, and it is unclear if aberrant regulation of complement underlies other symptoms of PNH. Currently available therapy for the hemolysis of PNH is suboptimal. Over the last decade, inhibitors have been developed that target the most critical functional sites of the complement cascade. These reagents have therapeutic potential in PNH. Ideally, the efficacy and safety of these inhibitors would be tested in an animal model of PNH to determine the optimal conditions and disease specific sites for complement inactivation. Due to differences in the mechanisms by which complement is regulated on murine cells, however, currently available transgenic mouse models of PNH are unlikely to be uninformative for these purposes. In the absence of an appropriate animal model, participation in phase I and phase II human trials appears to be the best available mechanism for developing a therapeutic strategy for complement inhibition in PNH. Careful analysis of the effects of these inhibitors will provide important insights into role of complement in the protean clinical manifestations of PNH.

Key words: hemoglobinuria, paroxysmal; complement; erythrocytes; hemolytic anemia

Hemolysis in PNH

Episodic hemoglobinuria is the clinical hallmark of PNH. Nocturnal hemoglobinuria was the symptom that defined PNH as an entity distinct from the two other hemoglobinurias (march hemoglobinuria and paroxysmal cold

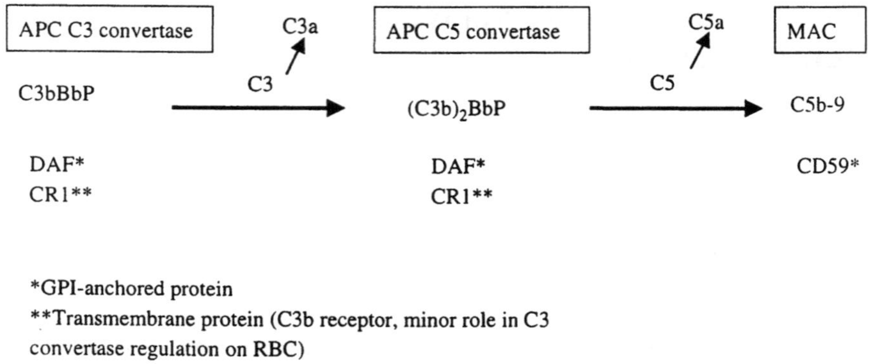

Figure 1. Regulation of the alternative pathway of complement by membrane inhibitors on human erythrocytes. Formation and stability of the C3 and C5 convertases are inhibited by both DAF and CR1 while MAC formation is regulated by MIRL (CD59). Although CR1 has both decay accelerating activity and factor I cofactor activity, it is present in a relatively low copy number (200-1000 molecules/cell) on RBC. Rather than an intracellular regulator of complement activation, it appears to function primarily as an intercellular receptor for C3 activation and degradation products that have been deposited on other surfaces or on immune complexes. CR1 is an integral membrane protein while DAF and MIRL are GPI anchored. Thus, DAF and MIRL are both deficient in PNH.

hemoglobinuria) also initially described in the late 19th century [1]. Studies aimed at defining the basis of this fascinating pathology dominated early investigation and led eventually to an understanding of the molecular basis of the disease.

About the hemolysis of PNH—what is certain?

In vitro, PNH erythrocytes are exquisitely sensitive to complement-mediated lysis [1]. This abnormal sensitivity is a consequence of combined deficiency of decay accelerating factor (DAF, CD55) and membrane inhibitor of reactive lysis (MIRL, CD59), the two proteins that normally regulate complement activity on the erythrocyte cell surface (Fig. 1).

About the hemolysis of PNH—what is likely?

Two lines of evidence suggest that aberrant regulation of complement accounts for the hemolytic anemia of PNH. First, Yonemura et al. [2] described a female patient with PNH who also had an inherited deficiency of complement C9. Despite having a very high percentage of the most complement sensitive PNH erythrocytes (95% of the patient's cells were DAF deficient), evidence of hemolysis was modest (serum LDH of 389 U/l, with the normal range being 130-250, and absence of hemosiderinuria). However, when the patient received whole blood transfusions following two separate surgical procedures, gross hemoglobinuria and scleral icterus were observed. In vitro, the patient's serum failed to mediate lysis of the patient's RBC, but addition of purified C9 restored hemolytic activity. These studies strongly support the hypothesis that the chronic hemolysis of PNH is complement-mediated.

Second, a patient with recurrent, episodic hemoglobinuria indistinguishable from that observed in patients with PNH was found to have an isolated deficiency of CD59 with normal expression of other glycosyl phosphatidylinositol-anchored proteins [3]. Subsequent analysis showed that the patient was homozygous for a deletional mutation that introduced a premature termination signal in CD59 mRNA [4]. Assuming that CD59 was the only abnormal RBC constituent, the recurrent episodes of intravascular hemolysis leading to hemoglobinuria must have resulted form failure to regulate the activity of the membrane attack complex (Fig. 1). Again, these findings are consistent with the concept that the hemolytic anemia of PNH is a consequence of aberrant regulation of complement because the only know function of erythrocyte CD59 is inhibition of membrane attack complex formation.

About the hemolysis of PNH—what is uncertain?

Although indirect, evidence that the hemolysis of PNH is complement-mediated is compelling [3, 5]. The mechanism by which complement is activated in vivo in a manner sufficient to mediate the chronic hemolysis of PNH, however, is largely speculative. Complement is in a state of continuous, low-level activation due to the "tick-over" phenomenon. This process results in activation of the alternative pathway of complement (APC) as a result of hydrolysis of the internal thioester bond of native C3 [6] (Fig. 2). The enzymatic complex that is generated by this process can initiate formation of the amplification C3 convertase of the APC. To form an efficient C5 convertase and to consequently generate the membrane attack complex (MAC, C5b-9), two molecules of activated C3 (C3b) must be bound to a particulate surface in close proximity. One C3b binds activated factor B, thereby forming the bimolecular enzymatic subunit of the C5 convertase (C3bBb). The other molecule of C3b binds C5 and positions it for cleavage by activated factor B. Because of the requirement for two C3b molecules to be bound in close proximity, it is unlikely that C5 convertases are formed in the fluid-phase as direct result of the tick-over phenomenon. Whether the bimolecular C3bBb complex under physiological conditions can activate C5 is speculative (discussed in more detail below).

Conceivably, the C5 convertase could form on the surface of PNH erythrocytes because DAF deficiency allows amplification of C3 activation. Two lines of

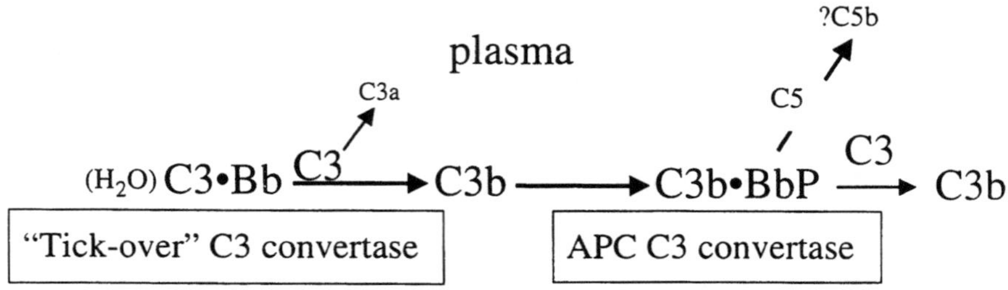

Figure 2. The tick-over phenomenon. The alternative pathway of complement (APC) is primed for activation at all times because the internal thioester bond is susceptible to hydrolysis. The conformational change induced by the process allows binding and subsequent activation of factor B by factor D, forming a functional C3 convertase. This enzymatic complex amplifies APC activation by converting native C3 to C3b thereby generating the nidus for formation of the physiological C3 convertase, C3bBb. Whether C5 can be activated in the fluid phase by a C3 convertase under normal physiological conditions is speculative.

evidence, however, suggest that formation of membrane associated C5 convertase is not the mechanism by which PNH erythrocytes undergo complement-mediated hemolysis in vivo.

First, erythrocytes from individuals with an isolated, inherited deficiency of DAF (the Inab phenotype) do not have activation products of C3 bound to the membrane [7]. This finding suggests that in vivo C3 convertase activity is normally inhibited even in the absence of DAF. Without initial formation of a C3 convertase, a C5 convertase is not generated. Thus, it is unlikely that the chronic hemolysis of PNH is a consequence of MAC formation mediated by cell-bound C3/C5 convertases.

Second, the patient described above with the isolated, inherited deficiency of CD59 had normal DAF expression [3]. Nonetheless the patient had recurrent episodes of intravascular hemolysis resulting in hemoglobinuria. Inasmuch as regulation of C3/C5 convertase activity on the patient's erythrocytes should be normal due to the presence of DAF, another mechanism for generating the cytolytic membrane attack complex is implied.

About the hemolysis of PNH-why does hemolysis increase during sleep?

Despite its importance in the history of PNH, the basis of the nocturnal exacerbations of hemoglobinuria remains enigmatic. In vitro, acidification of serum (optimum pH 6.4) activates the APC and induces lysis of PNH erythrocytes [1]. It is arguable, however, whether the very modest lowering of the pH that occurs during the relative hypoventilation of sleep is sufficient to enhance activation of the APC and thereby account for the diurnal variation in hemolysis in PNH. Perhaps there exists a "tick-over " phenomenon for C5b-9 generation analogous to that of C3 convertase formation and this process is affected by diurnally regulated factors. For example, some types of cobra venom factor (the snake analog of human C3c) [8] support C5 convertase formation [9]. In this case, a single molecule of cobra venom factor binds C5 with high affinity and positions it for cleavage by activated factor B that is bound to the same cobra venom factor molecule (CVF:Bb). Conceivably, some C3 molecules undergo conformational change that endows them with the capacity to bind both activated factor B and C5. Whatever the mechanism, the process must be enhanced during sleep. Understanding the basis of the nocturnal exacerbations of hemoglobinuria may provide new insights into basis of the chronic hemolysis of PNH and suggest a therapeutic approach to this troubling symptom.

Does complement activation cause manifestations of PNH other than hemolysis?

There is little doubt that the hemolysis of PNH is complement-mediated. Conceivably, other clinical manifestations of PNH are also a consequence of complement activation. For example, dysphagia is associated with PNH, and both the on-set and duration of this symptom correlate with hemolytic exacerbations [10]. In addition to the cytolytic MAC, complement activation generates cleavage products with important biological properties. Among these functionally important peptides are the C3a and C5a anaphylatoxins. Through receptor-ligand interactions, C3a and C5a mediate a variety of proinflammatory events (e. g., chemotaxis, smooth muscle contraction, vasodilatation, histamine release, and small vessel permeability, synthesis of IL-6 and TNF-alpha, and induction of the respiratory burst) [11]. Conceivably, the anaphylatoxins mediate the dysphagia of PNH by stimulating esophageal smooth muscle contraction. In vivo inhibition would illuminate the role of complement in the dysphagia of PNH.

Thromboembolic disease accounts for much of the morbidity and mortality associated with PNH [12]. There is some in vitro experimental support for the hypothesis that the thrombophilia of PNH is complement-induced [13]. In this case, CD59 deficiency results in increased complement-mediated vesiculation of platelets. The microvesicals generated as consequences of this process provide the phospholipid-binding site that serves as the nidus for formation of the prothrombinase complex. Whether this process is important in vivo is speculative.

Treatment of patients with specific inhibitors would provide an opportunity to investigate the role of complement in the thrombophilia of PNH.

Is aberrant regulation of complement simply an epiphenomenon?

PNH arises when a primitive hematopoietic stem cell in which *PIG-A* is somatically mutated undergoes clonal expansion. Consequently, all GPI-anchored proteins expressed on hematopoietic cells derived from the mutant clone are deficient. At least 17 GPI-anchored proteins are expressed on hematopoietic cells [14, 15]. Except for DAF (CD55) and MIRL (CD59), these proteins are functionally diverse, and the only pathology associated with PNH that has been shown to be directly linked to deficiency of GPI-AP is the hemolytic anemia.

The basis of the clonal expansion of *PIG-A* mutant stem cells in PNH is speculative. Deficiency of DAF (CD55) and MIRL (CD59) may be simply an epiphenomenon owing to their being GPI-anchored. Alternatively, aberrant regulation of complement may play a role in the clonal expansion of *PIG-A* mutant stem cells. In this case, deficiency of CD55 and CD59 would account for the growth or survival advantage of the mutant stem cell. Both CD55 and CD59 are abundantly expressed on $CD34^+$ $CD38^-$ hematopoietic stem cells [16]. Conceivably, products of complement activation stimulate hematopoiesis. According to this hypothesis, absence of the complement regulatory proteins results in greater complement activation that in turn enhances proliferation or survival of GPI-anchor deficient, *PIG-A* mutant stem cells.

Complement has a variety of non-immune functions that suggest mechanisms by which hematopoiesis could be affected. Complement C3 appears to be involved in limb regeneration in urodeles (salamander species) [17] and to stimulate B cell proliferation through complement receptor 2 [18]. Di Renzo and colleagues [19] reported that a hematopoietic progenitor cell line synthesized and cleaved C3. They showed that more differentiated cells acquired C3 binding sites, while transformed cells acquired plasma membrane proteases that activated the APC, resulting in binding of C3 cleavage products to the cell membrane. Recently, C5a was shown to be essential for normal liver regeneration in a murine model of CCl_4-indcued liver toxicity [20]. Sublytic amounts of CD59 appear to be mitogenic for fibroblasts [21] and aortic smooth muscle cells [22]. Together, these observations suggest that complement activation could influence hematopoiesis in an autocrine/paracrine fashion through binding of cleavage products to specific receptors (e. g., the C3a or C5a anaphylatoxin receptors or the CR1 or CR3 adhesion receptors). Alternatively, or in addition, binding of sublytic amounts of the MAC may be mitogenic for stem cells. Studies aimed at characterizing the role of complement in hematopoiesis are needed to investigate this hypothesis.

What should be the target of complement inhibition in PNH?

As discussed above, the mechanism of complement activation that accounts for the chronic hemolysis of PNH is incompletely understood. Broader knowledge of this process (and of the basis of the nocturnal exacerbations) would permit more rational targeting of a site for complement inhibition. In the absence of a complete understanding of the process, however, it is still possible to make an informed assessment about the effects of site-specific inhibition of the complement cascade on the hemolysis of PNH. Over the last decade, inhibitors that block the activity of all the critical sites of complement activation and complex formation have been developed [23]. Thus, reagents are available for analyzing the effects of complement inhibition in PNH.

Inhibition of the MAC

If the goal of therapy were to inhibit hemolysis only, targeting of the MAC would be appropriate, as direct inhibition of C5b-9 formation would not block the activity of the C3 and C5 convertases (Fig. 1). This result may be advantageous because all normal physiological functions of complement except cytolysis would remain functional. Alternatively, activation of the C3 and C5 convertases may have pathological consequence in PNH (e. g. the C3a and C5a anaphylatoxins may mediate dysphagia or MAC binding to platelets may contribute to thrombophilia). If so, inhibition of the MAC only would result in suboptimal therapy.

A number of reagents are available that block MAC formation. A recombinant, soluble form of CD59 (lacking the GPI-anchor) has been engineered and shown to inhibit complement-mediated lysis in vitro, although the specific activity was relatively low [24]. A chimeric CD59-antibody fusion protein was developed by Zhang and colleagues [25]. This type of reagent has therapeutic potential in PNH because the inhibitor could be targeted to a specific cell-type. For example, CD59-anti-glycophorin would be erythroid specific. The creation of a myristolated form of CD59 also has implication for treatment of PNH [26]. In this case, soluble CD59 (lacking the GPI-anchor) is derivatized with a myristolated peptide by disulfide exchange. This modification generates a molecule that retains aqueous solubility while acquiring the capacity to bind to the cell membrane. Potentially, this reagent would bind to circulating RBC (in addition to other blood cells) and inhibit the hemolytic anemia of PNH without interfering with the humoral functions of complement.

Blocking C5 convertase activity can also inhibit MAC formation (Fig. 1). A number of antibodies with excellent pharmacological characteristics have been developed [23]. One of these, a single chain, humanized, recombinant antibody has been evaluated in a Phase II clinical trail involving patients undergoing cardiopulmonary bypass [27]. At a dose of 2 mg/kg, the antibody inhibited complement activity by more than 50% for approximately 14 hours, and MAC formation during this period was not observed. Patients receiving the antibody had a significant reduction in myocardial damage and blood loss, and cognitive deficits were reduced.

Although the complement inhibitors described above appear to have therapeutic potential, recombinant proteins have several disadvantages in the clinical setting. Among these undesirable characteristics are cost, inconvenience (must be give intravenously), potential for antibody formation, and relatively short half-life. To overcome these limitations, efforts at developing small molecule inhibitors of complement have been initiated. In the case of C5 inhibition, Biesecker and colleagues [28] used combinatorial chemistry to develop an RNA aptamer inhibitor of C5. One of these aptamers bound C5 with a K_d of 2-5 nM and inhibited complement-mediated lysis of sheep erythrocytes. These results are encouraging, and in vivo characterization of this type of inhibitor is warranted.

Inhibition of the C3/C5 convertase

It might also be possible to ameliorate the hemolytic anemia of PNH by inhibiting the formation of the C3/C5 convertases. Recombinant, soluble CR1 is a remarkably effective inhibitor of complement activation [29]. This protein binds C3b and causes rapid decay of the enzymatic activity of the C3/C5 convertases of both the classical and alternative pathways of complement. In addition, it serves as a cofactor for factor I mediated cleavage of C3b into its inactive form iC3b. Recombinant CR1 has been modified and derivatized in a number of ways to enhance its therapeutic potential. A recombinant form specific for the alternative pathway, sCR1(desLHR-A) has been generated[30], and sCR1 glycosylated with sialyl Lewis x was shown to have bifunctional properties (inhibition of both complement activation and selectin-mediated adhesion) [31, 32]. Soluble CR1 has also been myristolated, a process that endows the derivatized protein with the capacity to bind to cell membranes [26].

Other recombinant proteins that target the C3/C5 convertase are soluble DAF (sDAF) [33] and soluble membrane cofactor protein (sMCP) [34]. A hybrid sDAF-sMCP molecule called complement activation blocker-2 (CAB-2) has also been developed [35]. As anticipated, CAB-2 has both decay accelerating factor and factor I cofactor activity. A DAF-CD59 hybrid has also been engineered [36]. Interestingly, this chimeric protein retained bifunctional activity only if the DAF portion were constructed amino-terminal to CD59 [36].

Compstatin, a 13-residue cyclic peptide isolated from a combinatorial peptide library, binds native C3 thereby inhibiting C3 convertase cleavage [37]. This inhibitor was found to have significant activity in ex vivo models of hyperacute rejection [38] and extracorporeal circulation [39]. Recent studies suggest activity in vivo as well [23].

The studies cited above represent an incomplete list of complement inhibitors that have been developed over the last decade. Nonetheless, from the above

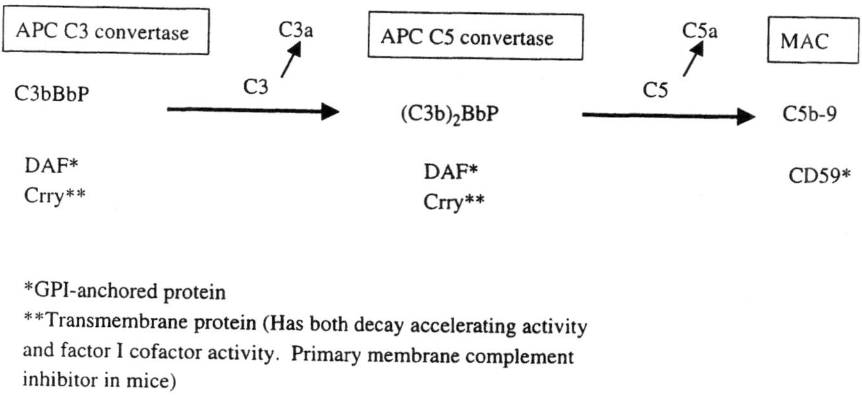

Figure 3. Regulation of the alternative pathway of complement by membrane inhibitors on mouse erythrocytes. As is the case for human RBC (Fig. 1), two proteins, one GPI-anchored (DAF) and one an integral membrane protein (Crry) inhibit both formation and stability of the C3 and C5 convertases on mouse erythrocytes. Like human CR1, mouse Crry has both decay accelerating activity and factor I cofactor activity. Unlike the case with CR1, however, Crry appears to be the primary regulator of complement on mouse erythrocytes. A murine ortholog of CD59 has been identified, but mice with a targeted deletion of CD59 have little evidence of spontaneous intravascular hemolysis, suggesting that MAC formation is controlled primarily by inhibition of C3/C5 convertase activity by Crry. Targeted disruption of Crry is incompatible with fetal viability.

discussion, it is clear that investigation in this area is vigorous and progressive. Ideally, a complement inhibitor with a relatively long half-life that can be taken orally will be developed. In the absence of such a reagent, sCR1 and scFv anti-C5 appear to be the best available candidates for therapy of the hemolysis of PNH. Both recombinant proteins are potent in vivo inhibitors of complement activation, and both are currently in clinical trials in humans.

Testing Complement Inhibitors in Animal models of PNH

An animal model of PNH could be used to identify inhibitors with favorable therapeutic indices. Germ line mutation of *PIG-A* is embryonic lethal in mice [40]. Using conditional knockout technology, mice mosaic for mutant and normal *PIG-A* have been generated [41]. Targeted gene disruption has also been used to create both CD55 [42] and CD59 [43] knockouts (CD59/CD55 double knocks have also been developed and are currently undergoing characterization). None of these animals, however, are good models for testing the efficacy of complement inhibitors as therapy for PNH because none have significant intravascular hemolysis. The difference between RBC susceptibility to complement in humans and mice is due, at least in part, to the presence of Crry on mouse erythrocytes (Fig. 3). This integral membrane protein has both decay accelerating and factor I cofactor activity [44], and it is the primary inhibitor of complement in mice [45]. Unlike CD55 and CD59 knockouts that have an essentially normal phenotype, targeted gene disruption of Crry results in loss of fetal viability. This embryonic lethality is complement-mediated because back breeding to C3 knockout mice restores viability of pups homozygous for mutant Crry. The marked difference in the way that complement is regulated on murine cells compared to human cells limits the value of mouse models of PNH.

Conclusions

In the absence of a good animal model of PNH, it seems reasonable to proceed with testing of complement inhibitors in patients with PNH. As noted above, two complement inhibitors are currently being tested in Phase II clinical trials. Candidates for therapy would be patients with symptomatic hemolysis, or patients experiencing an acute exacerbation of hemolysis. Careful study of treated patients is likely to provide new insights into the role of complement in the protean clinical manifestation of PNH.

References

1. Parker CJ. *Hemolysis in PNH.* In: Young NS, Moss, J., editor. PNH and the GPI-anchored proteins. San Diego: Academic Press; 2000. p. 49-100
2. Yonemura Y, Kawakita M, Koito A, Kawaguchi T, Nakakuma H, Kagimoto T, et al. Paroxysmal nocturnal haemoglobinuria with coexisting deficiency of the ninth component of complement: lack of massive haemolytic attack. *Br J Haematol.* 1990;74:108-113
3. Yamashina M, Ueda E, Kinoshita T, Takami T, Ojima A, Ono H, et al. Inherited complete deficiency of 20-kilodalton homologous restriction factor (CD59) as a cause of paroxysmal nocturnal hemoglobinuria. *N Engl J Med.* 1990;323:1184-1189
4. Motoyama N, Okada N, Yamashina M, Okada H. Paroxysmal nocturnal hemoglobinuria due to hereditary nucleotide deletion in the HRF20 (CD59) gene. *Eur J Immunol.* 1992;22:2669-2673
5. Yamaguchi M, Machii T, Azenishi Y, Nishimura J, Shibano M, Kanakura Y, et al. Detection of small populations of CD59-deficient erythrocytes in patients with aplastic anemia or myelodysplastic syndrome and normal individuals. *Blood Cells Mol Dis.* 2000;26:247-254

6. Pangburn MK, Schreiber RD, Muller-Eberhard HJ. Formation of the initial C3 convertase of the alternative complement pathway. Acquisition of C3b-like activities by spontaneous hydrolysis of the putative thioester in native C3. *J Exp Med.* 1981;154:856-867
7. Holguin MH, Martin CB, Bernshaw NJ, Parker CJ. Analysis of the effects of activation of the alternative pathway of complement on erythrocytes with an isolated deficiency of decay accelerating factor. *J Immunol.* 1992;148:498-502
8. Alper CA, Balavitch D. Cobra venom factor: evidence for its being altered cobra C3 (the third component of complement). *Science.* 1976;191:1275-1276
9. Rawal N, Pangburn MK. Functional role of the noncatalytic subunit of complement C5 convertase. *J Immunol.* 2000;164:1379-1385
10. Parker CJ, G. R. Lee. *Paroxysmal nocturnal hemoglobinuria.* In: Lee G. R. FJ, Greer J., Lukens L., Rodgers G., Paraskevas F., editor. Wintrobe's Clinical Hematology. 10th edition ed. Baltimore: Williams & Wilkens; 1998. p. 1264-1288
11. Drouin SM, Kildsgaard J, Haviland J, Zabner J, Jia HP, McCray PB, Jr., et al. Expression of the complement anaphylatoxin C3a and C5a receptors on bronchial epithelial and smooth muscle cells in models of sepsis and asthma. *J Immunol.* 2001;166:2025-2032
12. Sloand EMY, N. S. *Thrombotic Complications in PNH.* San Diego: Academic Press; 2000
13. Wiedmer T, Hall SE, Ortel TL, Kane WH, Rosse WF, Sims PJ. Complement-induced vesiculation and exposure of membrane prothrombinase sites in platelets of paroxysmal nocturnal hemoglobinuria. *Blood.* 1993;82:1192-1196
14. Kinoshita T, Inoue N, Takeda J. Defective glycosyl phosphatidylinositol anchor synthesis and paroxysmal nocturnal hemoglobinuria. *Adv Immunol.* 1995;60:57-103
15. Rosse WF. Paroxysmal nocturnal hemoglobinuria as a molecular disease. *Medicine (Baltimore).* 1997;76:63-93
16. Terstappen LW, Nguyen M, Huang S, Lazarus HM, Medof ME. Defective and normal haematopoietic stem cells in paroxysmal nocturnal haemoglobinuria. *Br J Haematol.* 1993;84:504-514
17. Del Rio-Tsonis K, Tsonis PA, Zarkadis IK, Tsagas AG, Lambris JD. Expression of the third component of complement, C3, in regenerating limb blastema cells of urodeles. *J Immunol.* 1998;161:6819-6824
18. Servis C, Lambris JD. C3 synthetic peptides support growth of human CR2-positive lymphoblastoid B cells. *J Immunol.* 1989;142:2207-2212
19. Di Renzo L, Zicari A, Longo A, Realacci M, Naso G, Pontieri G, et al. The hemopoietic progenitor 32DC13(G) cells synthesize C3 molecules and acquire C3b acceptor sites upon differentiation or neoplastic transformation. *J Immunol.* 1993;151:3737-3745
20. Mastellos D, Papadimitriou JC, Franchini S, Tsonis PA, Lambris JD. A novel role of complement: mice deficient in the fifth component of complement (C5) exhibit impaired liver regeneration. *J Immunol.* 2001;166:2479-2486
21. Halperin JA, Taratuska A, Nicholson-Weller A. Terminal complement complex C5b-9 stimulates mitogenesis in 3T3 cells. *J Clin Invest.* 1993;91:1974-1978
22. Niculescu F, Badea T, Rus H. Sublytic C5b-9 induces proliferation of human aortic smooth muscle cells: role of mitogen activated protein kinase and phosphatidylinositol 3-kinase. *Atherosclerosis.* 1999;142:47-56
23. Sahu A, Lambris JD. Complement inhibitors: a resurgent concept in anti-inflammatory therapeutics. *Immunopharmacology.* 2000;49:133-148

24.Sugita Y, Ito K, Shiozuka K, Suzuki H, Gushima H, Tomita M, et al. Recombinant soluble CD59 inhibits reactive haemolysis with complement. *Immunology*. 1994;82:34-41

25.Zhang HF, Yu J, Bajwa E, Morrison SL, Tomlinson S. Targeting of functional antibody-CD59 fusion proteins to a cell surface. *J Clin Invest*. 1999;103:55-61

26.Smith GP, Dodd, I., Davies, A., Morgan, B. P., Lachmann, P. J., Smith, R. A. G. Derivatization of soluble human CD59 with a myristoylated peptide creates a potent membrane-bound inhibitor of complement. *Immunopharmacology*. 2000;49:53

27.Fitch JC, Rollins S, Matis L, Alford B, Aranki S, Collard CD, et al. Pharmacology and biological efficacy of a recombinant, humanized, single-chain antibody C5 complement inhibitor in patients undergoing coronary artery bypass graft surgery with cardiopulmonary bypass. *Circulation*. 1999;100:2499-2506

28.Biesecker G, Dihel L, Enney K, Bendele RA. Derivation of RNA aptamer inhibitors of human complement C5. *Immunopharmacology*. 1999;42:219-230

29.Weisman HF, Bartow T, Leppo MK, Marsh HC, Jr., Carson GR, Concino MF, et al. Soluble human complement receptor type 1: in vivo inhibitor of complement suppressing post-ischemic myocardial inflammation and necrosis. *Science*. 1990;249:146-151

30.Murohara T, Guo JP, Delyani JA, Lefer AM. Cardioprotective effects of selective inhibition of the two complement activation pathways in myocardial ischemia and reperfusion injury. *Methods Find Exp Clin Pharmacol*. 1995;17:499-507

31.Huang J, Kim LJ, Mealey R, Marsh HC, Jr., Zhang Y, Tenner AJ, et al. Neuronal protection in stroke by an sLex-glycosylated complement inhibitory protein. *Science*. 1999;285:595-599

32.Rittershaus CW, Thomas LJ, Miller DP, Picard MD, Geoghegan-Barek KM, Scesney SM, et al. Recombinant glycoproteins that inhibit complement activation and also bind the selectin adhesion molecules. *J Biol Chem*. 1999;274:11237-11244

33.Moran P, Beasley H, Gorrell A, Martin E, Gribling P, Fuchs H, et al. Human recombinant soluble decay accelerating factor inhibits complement activation in vitro and in vivo. *J Immunol*. 1992;149:1736-1743

34.Christiansen D, Milland J, Thorley BR, McKenzie IF, Loveland BE. A functional analysis of recombinant soluble CD46 in vivo and a comparison with recombinant soluble forms of CD55 and CD35 in vitro. *Eur J Immunol*. 1996;26:578-585

35.Higgins PJ, Ko JL, Lobell R, Sardonini C, Alessi MK, Yeh CG. A soluble chimeric complement inhibitory protein that possesses both decay-accelerating and factor I cofactor activities. *J Immunol*. 1997;158:2872-2881

36.Fodor WL, Rollins SA, Guilmette ER, Setter E, Squinto SP. A novel bifunctional chimeric complement inhibitor that regulates C3 convertase and formation of the membrane attack complex. *J Immunol*. 1995;155:4135-4138

37.Sahu A, Kay BK, Lambris JD. Inhibition of human complement by a C3-binding peptide isolated from a phage-displayed random peptide library. *J Immunol*. 1996;157:884-891

38.Fiane AE, Mollnes TE, Videm V, Hovig T, Hogasen K, Mellbye OJ, et al. Compstatin, a peptide inhibitor of C3, prolongs survival of ex vivo perfused pig xenografts. *Xenotransplantation*. 1999;6:52-65

39.Nilsson B, Larsson R, Hong J, Elgue G, Ekdahl KN, Sahu A, et al. Compstatin inhibits complement and cellular activation in whole blood in two models of extracorporeal circulation. *Blood*. 1998;92:1661-1667

40.Kinoshita T, Bessler M., Takeda, J. *Animal models of PNH*. San Diego: Academic Press; 2000

41.Tremml G, Dominguez C, Rosti V, Zhang Z, Pandolfi PP, Keller P, et al. Increased sensitivity to complement and a decreased red blood cell life span in mice mosaic for a nonfunctional Piga gene. *Blood*. 1999;94:2945-2954

42.Sun X, Funk CD, Deng C, Sahu A, Lambris JD, Song WC. Role of decay-accelerating factor in regulating complement activation on the erythrocyte surface as revealed by gene targeting. *Proc Natl Acad Sci U S A*. 1999;96:628-633

43.Holt DS, Botto M, Bygrave AE, Hanna SM, Walport MJ, Morgan BP. Targeted deletion of the CD59 gene causes spontaneous intravascular hemolysis and hemoglobinuria. *Blood*. 2001;98:442-449

44.Kim YU, Kinoshita T, Molina H, Hourcade D, Seya T, Wagner LM, et al. Mouse complement regulatory protein Crry/p65 uses the specific mechanisms of both human decay-accelerating factor and membrane cofactor protein. *J Exp Med*. 1995;181:151-159

45.Xu C, Mao D, Holers VM, Palanca B, Cheng AM, Molina H. A critical role for murine complement regulator crry in fetomaternal tolerance. *Science*. 2000;287:498-501

Part 2

Pathogenesis and Animal Models of PNH

Mice with Blood Cells Deficient in GPI-linked Proteins – The Generation of a Murine Model to Investigate the Pathogenesis and Pathophysiology of Paroxysmal Nocturnal Hemoglobinuria (PNH)

Monica Bessler[1], Shashikant Kulkarni[1], Marek Jasinski[1], Anne Schaefer[1], Peter Keller[1] and Gabi Tremml[2,3]

[1]Division of Hematology, Internal Medicine, Washington University School of Medicine, 660 S. Euclid Ave, Campus Box 8125, St. Louis, 63110 MO, USA.
[2]Department of Human Genetics, Memorial Sloan Kettering Cancer Center, 1275 York Ave, Campus Box 110, New York 10021 NY, USA

Clinical characteristics of PNH include anemia due to intravascular hemolysis and hemoglobinuria, complications caused by bone marrow failure, and recurrent thrombosis. Despite the cloning of the *Piga* gene and the recent insights into the molecular lesion that causes the loss of glycosyl phosphatidylinositol-(GPI) linked proteins, much of the pathophysiology of this disease remains unexplained. A murine model that mimics the human disease would thus be a very valuable tool to study the pathogenesis and pathophysiology of PNH. Mice with blood cells deficient in GPI-linked proteins have been generated by several different approaches. Targeted recombination of the murine homologue *Piga* in embryonic stem cells (ES cells) is complicated by a high embryonic lethality and the lack of germline transmission of the disrupted *Piga* gene. To target *Piga* gene inactivation specifically to hematopoietic cells we and other investigators have therefore used conditional *Piga* gene inactivation employing the Cre/loxP system. In this review we will discuss Cre/loxP mediated *Piga* gene inactivation, summarize and compare the approaches that have been chosen to produce mice with PIGA(-) blood cells, and discuss the insights gained so far by analyzing mice with PIGA(-) blood cells. Further investigation of hematopoiesis and immune response in these mice might help to explain the pathogenesis and pathophysiology of PNH, provide new insights into the functional role of GPI-linked proteins on blood cells, and possibly provide the experimental system for studying PNH therapy.

Keywords: Mouse model, GPI-linked proteins on mouse blood cells, tissue specific *Piga* gene inactivation, Ham test in mice, hemolysis of murine PIGA(-) red blood cells, Cre/loxP

---

[3] Current address: Cell and Molecular Technologies, 580 Marshall Street, Phillipsburg, NJ 08865, USA.

Introduction

In the last seven years considerable progress in the field of PNH research has resulted from the cloning of the *PIGA* gene. *PIGA* encodes a protein essential in the biosynthesis of glycosyl phosphatidylinositol (GPI) anchor molecules. A mutation in this gene explains why a proportion of blood cells from patients with PNH is deficient in all GPI-linked surface proteins. Despite the recent insights into the molecular defects causing the blood cell phenotype, little is known about the specific pathways causing the human disease. The three cardinal clinical features of patients with PNH are anemia due to intravascular hemolysis and hemoglobinuria, complications of bone marrow failure, and an increased frequency of thrombosis. The mouse provides an excellent opportunity to study the pathogenesis and pathophysiology of the human disease. It is the ideal animal system because of the possibility of introducing specific mutations into the mouse genome, the availability of inbred mouse strains, the growing resources of mice with genetic mutations, the relative short generation time, the (relative) low cost of animal maintenance, and finally the knowledge of the mouse genome in the very near future. (N.B. In this review "*PIGA*" refers to the human gene, "*Piga*" to the murine gene, and "PIGA" to the gene product of either species [1]).

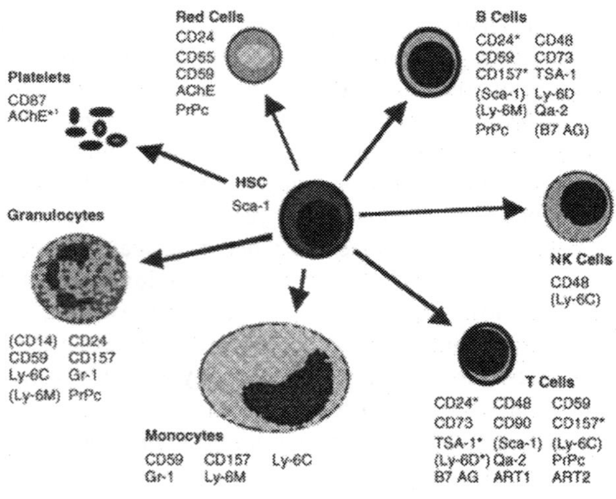

Fig. 1. GPI-linked proteins on murine blood cells. At least 21 different GPI-anchored surface proteins are known to be expressed on murine blood cells. We assume that CD55 and CD59 are expressed on all blood cells, although this has not been studied in detail.
"( )": Only expressed upon activation or only in a subset of cells. *: Only expressed during differentiation, but not on mature cells in peripheral blood. [1]: In megakaryocytes mainly in a granular, soluble forms; expression as a GPI-linked form unknown. AChE: acetylcholinesterase. PrPc: Prion protein. ART1 and 2: ecto-ADP-ribosyltransferases.

Mice, like humans, carry a number of surface proteins on blood cells that are linked to the membrane by a GPI-anchor molecule. Some of these molecules are the same or similar to those on human red cells. Others have either not been found on human blood cells, for example Sca-1, or have a different lineage distribution, for example CD24, which while expressed on almost all murine blood cells, in humans is only found on B-cells. Figure 1 lists some of the GPI-linked molecules and their lineage distribution on murine blood cells.

Piga Gene Inactivation in Murine ES Cells

Obeying the evolutionary conservation of synteny of the mammalian X chromosome ("Ohno's law" [2]) the murine *Piga* maps to X-F3/4 and is highly homologous to the human gene [3]. In vitro differentiation of ES cells with an inactive *Piga* gene {PIGA(-) ES cells} demonstrated that hematopoietic differentiation occurs in the absence of GPI-anchored molecules [4, 5]. However, injection of PIGA(-) ES cells into blastocyst was embryonic lethal when a high contribution of PIGA(-) cells was obtained. Only low percentage chimeras survived to adulthood. Surviving chimeric mice had a small proportion of PIGA(-) blood cells. No germline transmission of the mutant *Piga* gene was observed [5-7].

Injection of PIGA(-) ES cells into blastocysts from RAG(-) mice lead to the full restoration of the B and T cell compartment in RAG(-) animals. Interestingly however, innate immunity in mice with PIGA(-) lymphocytes was severely disturbed, suggesting that GPI-linked proteins are important for maintaining peripheral lymphoid homeostasis [8].

Conditional *Piga* Gene Inactivation

Because *Piga* null mutations are not compatible with life [5-7] the Cre/lox recombination system was employed. The Cre enzyme, isolated from the bacteriophage P1 belongs to the family of integrases, and catalyzes the recombination event between two palindromic l*oxP* consensus sequences [9]. Cre binds to each palindromic strand of two *loxP* sites, and brings them together by forming a tetramer. Recombination occurs in a small spacer area between the palindromes [10, 11]. The Cre/lox system is being used in inducible gene targeting experiments [12] (reviewed in Sauer 1998 [13]), in tissue specific gene targeting experiments [14] (reviewed in Nagy 1999 [15]), and in inducible/tissue specific gene targeting experiments, where temporal and spatial specificity is provided by a heterologous promoter [16-18]. We and other investigators have developed mice in which Cre mediated DNA excision inactivates the *Piga* gene [19, 20]. The strategy to introduce the *loxP* sites into the *Piga* locus used by Tremml et al. [20] is illustrated in Fig. 2.

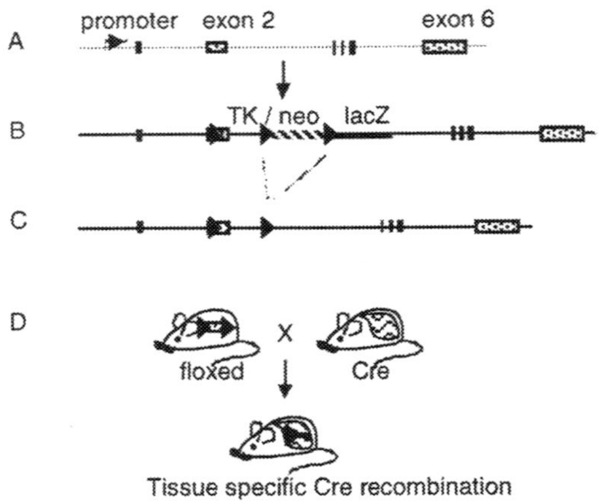

Fig 2. Conditional inactivation of *Piga*. A-C demonstrates the strategy to introduce the loxP sequences into the *Piga* locus as done by Tremml et al. [20]. D schematically illustrates Cre mediated inactivation of PIGA by genetic breeding. A) Genomic organization of the *Piga* locus shows exons in shaded boxes. B) Modified *Piga* locus after homologous recombination in ES cells: One *loxP* site (arrows) is inserted upstream of the initiation ATG, in the same frame as the *lacZ* (black box) coding region. The two downstream *loxP* sites flank the hsvTK/pgk-neo selection cassette (dark shaded box). The ***neoR*** gene was used to select for the presence of homologous recombination in ES cells. HsvTK selection using Gancyclovir was used to screen for excision of the selection cassette. C) Homologous recombined ES cells were subjected in vitro to Cre mediated excision of the *hsvTK/pgk-neo* selection cassette. ES cells with partial site specific recombination (shown in stippled lines) and two *loxP* sites left in the *Piga* locus were selected (*loxPiga*). In this configuration the function of the *Piga* gene is not impaired and lacZ is not expressed. D) ES cells with two *loxP* sites and *lacZ*are injected into blastocyst for the generation of *loxPiga* mice. These in turn are bred to transgenic mice carrying a *cre* gene under the desired tissue specific promoter. In the progeny carrying both the floxed *Piga* locus and the *cre* transgene, tissue specific recombination of the *loxPiga* allele occurs creating a fusion gene between *Piga* and *lacZ*. Thus, cells with a deleted *Piga* gene will not only lack GPI-linked proteins but will also express LacZ driven by the endogenous *Piga* promoter.

Mice of the floxed strain (*loxPiga* mice) are bred to transgenic mice carrying Cre recombinase under the control of a tissue specific promoter. Because the progeny will carry the floxed *Piga* gene as well as the *cre* transgene, *Piga* will be excised due to Cre/lox recombination exclusively in cells where Cre is active. Once *Piga* is excised in a cell, this genetic alteration will be inherited in all of the descendant cells. Mice from a floxed *Piga* strain can thus be used to study the role of GPI-linked proteins in a variety of tissues, depending on the expression of Cre. *Piga* is subject to X chromosome inactivation [21]. Thus in males, the inactivation of the one *Piga* allele is sufficient to cause the loss of GPI-linked proteins on the cell surface. In females, the inactivation of the *Piga* allele only when present on the active X-chromosome will lead to the loss of GPI-linked proteins on the cell surface. In females heterozygous for a floxed *Piga* allele, due to random X-inactivation, only about half of the cells will lack GPI-linked proteins.

Mice with a floxed *Piga* allele generated by Tarutani and colleagues [19] differ from our loxPiga animals in that a) Tarutani chose exon 6 to be flanked by loxP sites. Exon 6 of *Piga* encodes the transmembrane domain. Its excision results in a truncated transcript and the failure to localize PIGA to the endoplasmic reticulum in order to exert its function. b) Tarutani inserted a neoR selection cassette 3' of the *Piga* gene, which was not removed but remained in the floxed *Piga* mouse strain. Although there are reports of hypomorphic phenotypes caused be the presence of the neoR gene [22, 23], its presence seems not to alter the normal PIGA function. c) No positive marker gene was inserted that positively identifies PIGA(-) cells (LacZ in Tremml et al.[20]).

Mice with PIGA(-) Blood Cells Generated by Conditional Inactivation of *Piga*

Piga Gene Deletion in the Early Embryo

Mice with PIGA(-) blood cells were obtained by various strategies. In a first approach mice mosaic for PIGA(-) cells were produced. We expressed Cre under the adenoviral *EIIa* promoter resulting in *Piga* deletion during a short time period in early embryogenesis [20]. Offspring were mosaic for cells with an inactive PIGA. The extent of mosaicism was dependent on the efficiency of *Piga* gene deletion in early embryogenesis and in female mice the balance of X-chromosome inactivation [20, 21] (Fig. 3C and 3D). Murakami and colleagues [24] used Cre expression under a human CMV promoter, which results in ubiquitous and continuous *Piga* gene deletion, leading to multiple developmental abnormalities and embryonic lethality [25]. Thus, fetal liver cells from female embryos were transplanted into lethally irradiated hosts which gave rise to hematopoiesis mosaic for PIGA(-) blood cells. In these mice the extent of blood cell mosaicism was dependent on the donor chimerism and on the balance of X-inactivation in the donor cells [24] (Figure 3A and 3B). With the exception of the T cells in both mouse models, the proportions of GPI(-) blood cells remained unchanged during an observation period of up to 18 month, indicating that under normal conditions PIGA(-) hematopoietic progenitor cells do not have a growth advantage. Interestingly, in both mouse models the number of PIGA(-) T cells increased. This was accentuated in mice with PIGA(-) blood cells obtained after transplantation of PIGA(-) fetal liver cells but was also found in the mosaic EIIa-cre derived mice, although to a lesser extent. These results suggest an impaired homeostasis of

PIGA(-) T-cells. However, neither an increased incidence of leukemia nor signs of bone marrow failure were observed in these mice.

In mice mosaic for PIGA(-) cells germline transmission of the mutated *Piga* allele was not achieved, indicating that a *Piga* null mutation in mice and probably likewise in humans, is not heritable [21].

Although both mouse models nicely mimic the blood cell phenotype seen in patients with PNH they have certain shortcomings. Mosaic mice obtained by Tremml et al. are mosaic also in non-hematopoietic organs which might alter hematopoiesis. The model generated by Murakami et al. uses irradiated recipient mice. Gamma irradiation alters the microenvironment, which in turn might alter PIGA(+) and PIGA(-) hematopoiesis.

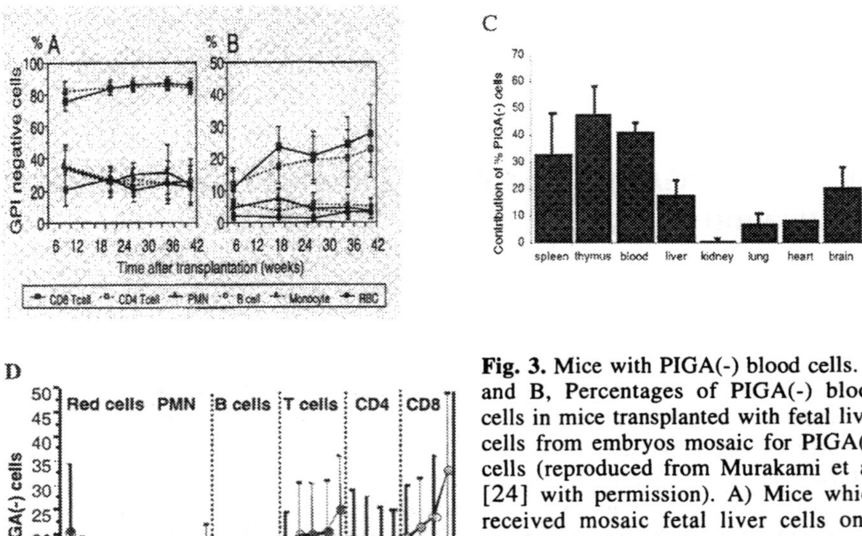

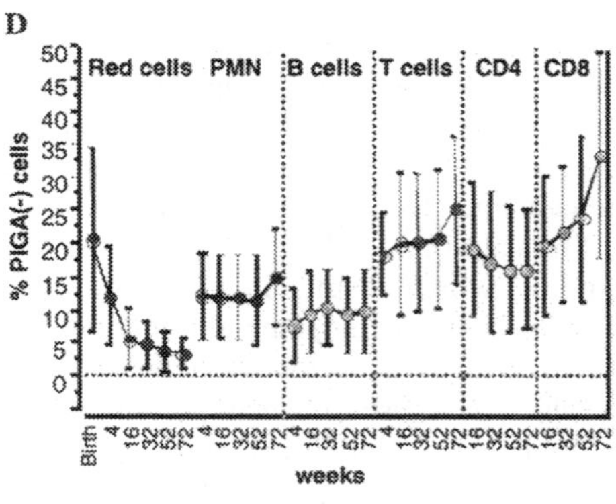

Fig. 3. Mice with PIGA(-) blood cells. A and B, Percentages of PIGA(-) blood cells in mice transplanted with fetal liver cells from embryos mosaic for PIGA(-) cells (reproduced from Murakami et al. [24] with permission). A) Mice which received mosaic fetal liver cells only (n=6); B) Mice which received a mixture of mosaic and normal fetal liver cells (n=22); C) Contribution of PIGA(-) cells to various organs in mice mosaic for PIGA(-) cells [21]; D) Proportion of PIGA(-) blood cells in various hematopoietic lineages of PIGA(-) mosaic mice (n=13) [20]. Bars indicate standard deviation.

Piga gene Inactivation Targeted to Hematopoietic Cells

Three different promoters driving the expression of Cre have been used to target *Piga* gene deletion to hematopoietic cells [26-28] (Fig. 4). Takahama and colleagues used the $p56^{lck}$ proximal promoter, which causes transgene expression in T cells commencing during thymic development. PIGA inactivation in the T cell compartment occurred to almost 100% (Fig. 4D). PIGA(-) T cells upon stimulation showed a normal expression of activation antigens and displayed adequate IL-2 secretion, cell cycle induction and cell proliferation, suggesting that a functional T cell pool can exist in the absence of GPI-linked surface proteins.

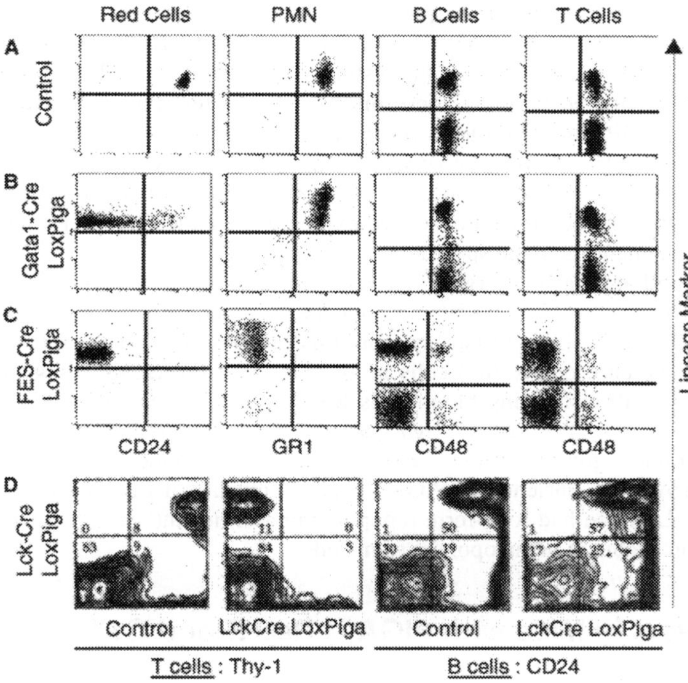

Fig. 4. Cre mediated inactivation of PIGA in specific blood cell lineages. Mice with a floxed *Piga* gene [20, 26] were crossbred with different Cre transgenic mice. Peripheral blood cells (first three panels) or splenocytes (bottom panel) were analyzed by flow cytometry. A) Expression of CD24, Gr-1, and CD48 (all GPI-linked) in a normal control mouse. B) Cre expression under transcriptional control of Gata1 with *Piga* inactivation specifically in the erythro/megakaryocytic lineage [29]. C) Cre expression under the transcriptional control of human *c-FES* with *Piga* inactivation in the hematopoietic stem cell and PIGA(-) cells in all blood cell lineages [28]. D) T cell specific *Piga* inactivation using Cre expression under the proximal $p56^{lck}$ promoter [26] (reproduced from Takahama et al. [26] with permission). The GPI-linked marker molecules are displayed on the X-axis and are indicated below the dot blot. The markers used for lineage determination are displayed on the Y-axis and were CD11b for granulocytes, B220 for B cells, and TcR for T cells. Red blood cells were identified by light scatter characteristics. PMN: polymorphonuclear cells.

We targeted the inactivation of PIGA to the erythroid compartment by driving Cre expression under the control of the DNA sequences required for erythroid/megakaryocytic expression of the murine GATA-1 transcription factor[29]. This breeding was associated with high embryonic lethality due to a leaky expression of Cre in very early embryogenesis. However, about 30% of mice that carried a *loxPiga* and a *cre* transgene (GL mice) escaped early recombination and were born alive and well. All GL animals born alive had PIGA(-) red blood cells, megakaryocytes, mast cells and eosinophils, corresponding to the expression pattern of the GATA-1 gene (Fig. 4B). In male GL mice or female mice homozygous for *loxPiga* up to 100% of red cells were deficient in GPI-linked proteins. Interestingly, in addition to red cells completely devoid of GPI-linked proteins, these mice also exhibited red cells with a partial deficiency of GPI-linked proteins resembling the PNH type II cells identified in patients with PNH. No or only very low numbers of PIGA(-) cells were found in the myeloid and lymphoid blood cell lineage and in non-hematopoietic organs, indicating that in the adult mouse Cre expression by the GATA-1 transcription regulatory elements is tissue specific.

In a second approach to target *Piga* gene inactivation to the early hematopoietic progenitor cells, we expressed Cre under the locus control region (LCR) of the human *c-FES* gene [28]. Offspring from this breeding (LF mice) showed a time dependent increase of PIGA(-) cells in all blood cell lineages leading to mice in whom almost all circulating blood cells were completely deficient in GPI-linked proteins (Fig. 4C). The rate of *Piga* gene inactivation increased when mice were bred for homozygousity of the *c-FES* transgene (LFF mice). Transplantation of bone marrow cells enriched in PIGA(-) cells showed that PIGA(-) bone marrow cells from LF mice had long-term repopulating abilities indicating that *Piga* gene inactivation occurs in hematopoietic stem cells.

Hemolysis in Mice with PIGA(-) Red Blood Cells

Anemia, obvious hemoglobinuria and signs of thrombosis were not observed in mice with PIGA(-) blood cells. However, mice with a high proportion of PIGA(-) red blood cells had decreased hemoglobin levels when compared with sex, strain, and age matched control mice (Fig. 5A). Likewise their reticulocyte counts were increased (Fig. 5B). In mice with PIGA(-) red blood cells, the proportion of PIGA(-) reticulocytes was always higher than the proportion of PIGA(-) red cells, indicating a reduced half-life of PIGA(-) red cells in circulation. Analogous to the human disease the reduced half-life of PIGA(-) red cells is most likely caused by an increased sensitivity of red cells to complement-mediated lysis. Indeed, a Ham test adapted for use in mice demonstrated lysis of PIGA(-) red blood cells when exposed to complement activated serum at dilutions that did not affect red blood cells derived from normal control mice [20, 24, 27] (Fig. 6). Red cells from GL mice that only had a partial deficiency of GPI-linked proteins showed an increased sensitivity towards complement compared to normal murine red blood cells as well. However, hemolysis in GL mice was less when compared to red blood cells that are completely devoid of GPI-linked proteins completely, consistent with their imitation of the PNH type II phenotype.

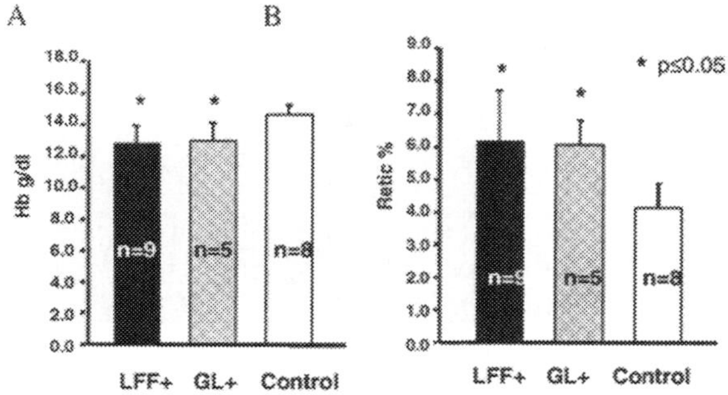

Fig 5. Hemoglobin values (A) and reticulocyte counts (B) in male LFF mice, male GL mice and normal male control mice. * indicates statistically significant difference compared to normal control mice.

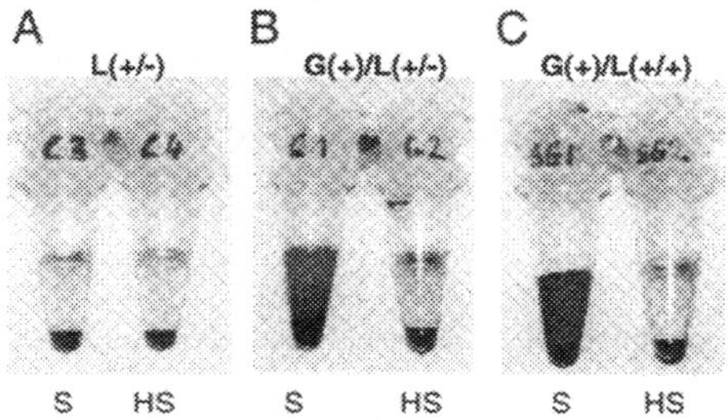

Fig 6. Ham test with red blood cells obtained form GL mice. No hemolysis is seen in normal control mice (A). In contrast, increasing hemolysis is seen in mice with 50% of red cells deficient in GPI-linked proteins (B) or 100% of red cells deficient in GPI-linked proteins (C). Diluted human serum served as a source of complement (S), heat inactivated serum was used as a control (HS).

Patients with the hemolytic form of PNH show a hypercellular bone marrow with a marked increase in erythropoiesis. In mice with PIGA(-) red blood cells, the bone marrow looks almost undistinguishable from that of normal control mice (Fig. 7A and B). In contrast, in spleens of mice with almost 100% of PIGA(-) red cells erythropoiesis is dramatically increased. This is consistent with a compensatory accelerated erythropoiesis and the spleen being a major erythropoietic organ in mice particularly when there is increased peripheral destruction [30].

In comparison to humans with PNH the half life of PIGA(-) red cells in mice is only moderately reduced and it remains to be shown whether this is due to intravascular hemolysis or rather to a premature removal of mature PIGA(-) red cells by the reticulo-endothelial system. A possible explanation for the rather mild anemia even in mice with 100% of PIGA(-) blood cells might be that in mice Crry, a transmembrane complement regulatory molecule, protects murine red cells against the action of complement.

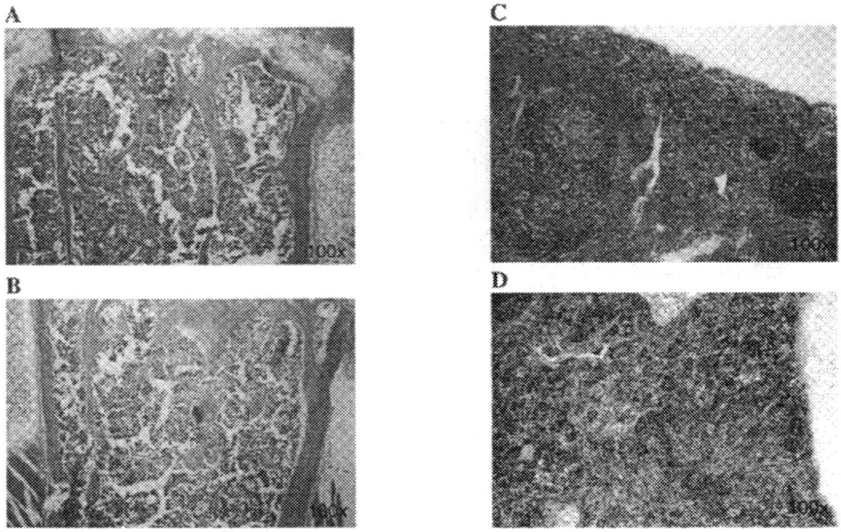

Fig 7. Erythropoiesis in bone marrow and spleen from a normal mouse and a GL+ mouse. A) Bone marrow from a normal control mouse. B) Bone marrow from a GL mouse. C) Spleen from a normal control mouse. D) Spleen from a GL mouse. Note, no increased erythropoiesis is seen in the bone marrow of GL mice, whereas erythropoiesis in the spleen is dramatically increased. Sections were stained with Hematoxylin and Eosin.

Insights Gained from Mice with PIGA(-) Blood Cells

In mice, as in humans, the loss of PIGA function causes the loss of GPI-linked proteins on the surface of cells. Breeding experiments with chimeric male mice and female mice mosaic for PIGA(-) cells suggested that a *Piga* null-mutation is not heritable. In fact, there are no reports of an inherited form of PNH in humans, in which cells lack GPI-linked proteins completely. In mice, as in humans, PIGA(-) hematopoietic cells contribute to all blood cell lineages. However the proportion of PIGA(-) cells varies between different blood cell lineages, suggesting that the lack of GPI-linked proteins has subtle effects on maturation or homeostasis within the individual blood cell compartment.

A *Piga* gene mutation alone in an otherwise neutral environment does not cause the expansion of PIGA(-) progenitor cells in our mice. This suggests that clonal expansion of the abnormal cells in patients with PNH is caused by a mechanism that selects for cells that lack GPI-linked proteins. Similarly, mice with PIGA(-) blood cells do not develop cytopenia, suggesting that the PNH clone is not responsible for bone marrow failure in patients with PNH.

In mice the proportion of PIGA(-) cells is very stable, with the possible exception of PIGA(-) T cells, which slowly increase in aging mice. This suggests that PNH hematopoiesis once established is unlikely to disappear. This is in accordance with the observation that in the majority of patients the relative size of the PNH clone is very stable.

Future Directions

Many features of PNH remain unexplained, for example, the pathogenesis of thrombosis or the cause of clonal expansion. Mice with PIGA(-) blood cells promise to be a valuable tool to address these questions. Naturally, there are limits to this model, because mice are not men. However, the possibility to exploit various genetic backgrounds and the possibility to apply specific experimental conditions may enable us to more closely imitate the situation prevailing in PNH. This may lead to valuable insights of the functional role of certain GPI-linked proteins that when absent might explain certain features of PNH.
Finally the availability of a mouse model for PNH will be valuable for the testing of new agents to cure the disease [31] or to more efficiently treat complications such as thrombosis or hemolysis.

References

1. Davisson M T (1997) Rules and guidelines for genetic nomenclature in mice: excerpted version. Committee on Standardized Genetic Nomenclature for Mice. Transgenic Res 6: pp 309-319
2. Ohno S (1967) Sex chromosomes and sex-linked genes, in Monographs on endocrinology, Labhart A, et al, Editors. Springer-Verlag: Berlin Heidelberg New York. pp 46-69
3. Kawagoe K, et al (1994) Molecular Cloning of murine Pig-a, a gene for GPI-anchor biosythesis, and demonstration of interspecies conservation of its structure, function, and genetic locus. Genomics 23: pp 566-574
4. Dunn D E, et al (1996) A knock-out model of paroxysmal nocturnal hemoglobinuria: Pig-a(-) hematopoiesis is reconstituted following intercellular transfer of GPI- anchored proteins. Proc Natl Acad Sci U S A 93: pp 7938-7943
5. Rosti V, et al (1997) Embryonic stem cells without pig-a gene activity are competent for hematopoiesis with the PNH phenotype but not for clonal expansion. J Clin Invest 100: pp 1028-1036
6. Kawagoe K, et al (1996) Glycosylphosphatidylinositol-anchor-deficient mice: Implications for clonal dominance of mutant cells in paroxysmal nocturnal hemoglobinuria. Blood 87: pp 3600-3606
7. Lin S R, et al (2000) Chimaeric mice with disruption of the gene coding for phosphatidylinositol glycan class A (Pig-a) were defective in embryogenesis and spermatogenesis. Br J Haematol 110: pp 682-693.
8. Rosti V, et al (1997) PNH-like lymphocytes are competent for generating the entire immune system in rag null mice. Blood 90: pp 159a (abs 698)

9. Sauer B and Henderson N (1988) Site specific DNA recombination in mammalian cells by the Cre recombinase of bacteriophage P1. Proc Natl Acad Sci U.S.A. 85: pp 5166-5170
10. Guo F, Gopaul D N, and van Duyne G D (1997) Structure of Cre recombinase complexed with DNA in a site-specific recombination synapse. Nature 389: pp 40-46.
11. Gopaul D N, Guo F, and Van Duyne G D (1998) Structure of the Holliday junction intermediate in Cre-loxP site- specific recombination. Embo J 17: pp 4175-4187.
12. Lakso M, et al (1992) Targeted oncogene activation by site-specific recombination in transgenic mice. Proc Natl Acad Sci U S A 89: pp 6232-6326
13. Sauer B (1998) Inducible gene targeting in mice using the Cre/lox system. Methods 14: pp 381-392.
14. Gu H, Zou Y R, and Rajewsky K (1993) Independent control of immunoglobulin switch recombination at individual switch regions evidenced through Cre-loxP-mediated gene targeting. Cell 73: pp 1155-1164.
15. Nagy A and Rossant J (1996) Targeted mutagenesis: analysis of phenotype without germ line transmission. J Clin Invest 97: pp 1360-1365.
16. Kuhn R, et al (1995) Inducible gene targeting in mice. Science 269: pp 1427-1429
17. Feil R, et al (1996) Ligand-activated site-specific recombination in mice. Proc Natl Acad Sci U S A 93: pp 10887-10890
18. St-Onge L, Furth P A, and Gruss P (1996) Temporal control of the Cre recombinase in transgenic mice by a tetracycline responsive promoter. Nucleic Acids Res 24: pp 3875-3877.
19. Tarutani M, et al (1997) Tissue specific knockout of the mouse pig-a gene reveals important roles for GPI-anchored proteins in skin development. Proc Natl Acad Sci USA 94: pp 7400-7405
20. Tremml G, et al (1999) Increased sensitivity to complement and a decreased red cell life span in mice mosaic for a non-functional Piga gene. Blood 94
21. Keller P, et al (1999) X inactivation and somatic cell selection rescues female mice carrying a Piga-null mutation. Proc Natl Acad Sci 96: pp 7479-7483
22. Meyers E N, Lewandoski M, and Martin G R (1998) An Fgf8 mutant allelic series generated by Cre- and Flp-mediated recombination. Nat Genet 18: pp 136-141.
23. Nagy A, et al (1998) Dissecting the role of N-myc in development using a single targeting vector to generate a series of alleles. Curr Biol 8: pp 661-664.
24. Murakami Y, et al (1999) Different roles of glycosylphosphatidylinositol in various hematopoietic cells as revealed by a mouse model of paroxysmal nocturnal hemoglobinuria [see comments]. Blood 94: pp 2963-2970
25. Nozaki M, et al (1999) Developmental abnormalities of glycosylphosphatidylinositol-anchor- deficient embryos revealed by Cre/loxP system. Lab Invest 79: pp 293-299
26. Takahama Y, et al (1998) Functional competence of T cells in the absence of glycosylphosphatidylinositol-anchored proteins caused by T cell- specific disruption of the Pig-a gene. Eur J Immunol 28: pp 2159-66
27. Jasinski M, et al (2000) GATA1-Cre mediates Piga gene inactivation in the erythroid/megakarycytic lineage and leads to circulating RBC's with a partial deficiency of GPI-linked proteins (PNH TYPE II cells). Blood 96: pp abs 2261
28. Keller P, et al (2001) FES-Cre targets PIGA inactivation to hematopoietic stem cells in the bone marrow. J Exp Med: pp In press

29. Jasinski M, et al (2001) GATA1-Cre mediates Piga gene inactivation in the erythroid/megakaryocytic lineage and leads to circulating red cells with a partial deficiency in GPI-linked proteins (PNH type II cells). Blood: pp In press
30. Hara H and Ogawa M (1976) Erthropoietic precursors in mice with phenylhydrazine-induced anemia. Am J Hematol 1: pp 453-458
31. Nishimura J, et al (2001) Efficient retrovirus-mediated PIG-A gene transfer and stable restoration of GPI-anchored protein expression in cells with the PNH phenotype. Blood 97: pp 3004-3010.

A Multistep Model for the Pathogenesis and Evolution of PNH

Russell E. Ware MD PhD[1], Matthew M. Heeney MD[1], Chrisley V. Pickens[1], Carlos M. DeCastro MD[2], and Thad A. Howard[1]

[1]Division of Hematology/Oncology, Pediatrics, Duke University Medical Center, Durham NC 27710, USA
[2]Division of Medical Oncology, Internal Medicine, Duke University Medical Center, Durham NC 27710, USA

Summary: Although paroxysmal nocturnal hemoglobinuria (PNH) has been recognized as an intriguing clinical entity for over a century, the past 10-15 years have witnessed remarkable progress toward a better understanding of the pathogenesis and pathophysiology of this mysterious blood disorder. After multiple proteins were found to be absent from the surface of PNH cells, it was recognized that each utilized a glycosylphosphatidylinositol (GPI) anchor for cell surface attachment. Elucidation of the biochemical abnormalities in PNH cells identified a specific defective step in GPI anchor bioassembly, which eventually led to the cloning of the PIG-A gene. All patients with PNH reported to date have acquired (non-germline) PIG-A gene mutations.

Despite this improved understanding of the biochemical and molecular abnormalities found in PNH, many facets of this disorder remain unexplained. The importance of the PIG-A mutations in PNH, the frequent association between PNH and aplastic anemia, the mechanisms for PNH cell growth and selection, and the transformation of PNH into myelodysplasia or leukemia are just several aspects of PNH that are poorly understood. Previous models for the development of PNH are somewhat simplistic and do not address each of these important questions. In this report, we suggest a new multistep model for the pathogenesis and evolution of PNH.

Key words: Aplastic anemia, Apoptosis, PIG-A gene mutations, PNH clones

Introduction

Paroxysmal nocturnal hemoglobinuria (PNH) is a fascinating hematological disorder with various clinical manifestations including episodic hemolysis (from which the name derives), venous thrombosis, and deficient hematopoiesis [1,2]. Occasionally, PNH will evolve into other related clinical syndromes such as aplastic anemia (AA), myelodysplasia (MDS), or acute leukemia. PNH is a clonal disorder, and typically cells within the PNH clone co-exist with normal hematopoietic cells in the bone marrow and peripheral blood. A key feature of PNH cells is the absence of proteins that utilize a glycosylphosphatidylinositol (GPI) anchor for attachment to the cell surface [3,4]. GPI anchor biosynthesis is defective in PNH at an early step [5,6], so completed anchors are not available for GPI-linkage. The molecular defect in PNH is an acquired mutation in the X-linked phosphatidylinositol glycan Class A (PIG-A) gene [7,8]; all reported patients with PNH have a PIG-A gene mutation [9-12].

Despite this increased level of understanding of the clinical, biochemical, and molecular abnormalities found in PNH, many important questions remain unanswered. What is the etiology of the PIG-A mutation in PNH, which is presumed to be an acquired mutation within a hematopoietic stem cell? Is the PIG-A mutation necessary or sufficient for the development of PNH? Do normal persons have PIG-A mutant hematopoietic cells? What is the nature of the association between PNH and aplastic anemia? By what mechanisms are the PIG-A mutant hematopoietic cells selected for growth, leading to clonal expansion and sometimes clonal dominance? Why do some cases of PNH evolve to aplasia or transform into myelodysplasia or leukemia? To help answer these questions, several models have been developed that attempt to incorporate the available clinical and laboratory observations for patients with PNH.

Previous Models of the Pathogenesis of PNH

Although PNH is a complex clinical disorder, the identification of PIG-A gene mutations in all reported cases suggests that the presence of this mutation plays a critical role in the pathogenesis of PNH. Two models have been previously proposed for the development of PNH. Both include the PIG-A gene mutation as a central event, in fact the *sine qua non* for the pathogenesis of PNH, but differ in their view of whether or not the PIG-A gene mutation leads directly to PNH.

One-step model

This simple model posits that the PIG-A mutation is both necessary and sufficient for the development of PNH. The PIG-A gene mutation provides an intrinsic growth advantage leading to clonal expansion and PNH (Figure 1).

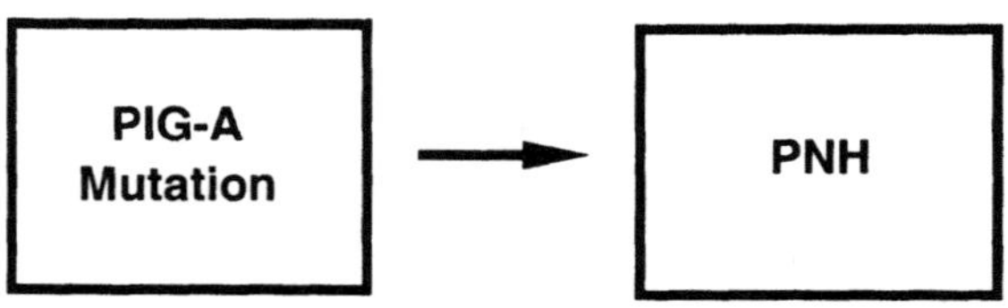

Fig. 1. One-step model for the development of (PNH). A hematopoietic stem cell acquires a PIG-A mutation, presumably as a random event. This mutation is both necessary and sufficient for clonal expansion and the eventual development of PNH.

The one-step model is attractive because of its simplicity and its conformity with the "one gene, one disease" theory. Clinical and laboratory observations, however, have not always fit this model. For example, some persons with relatively small numbers of GPI-deficient peripheral blood cells (e.g., 1-10%), maintain this clonal population in a stable fashion for many years without inexorably evolving toward clonal dominance. Furthermore, the identification of PNH patients with multiple clones, each with its own distinct PIG-A gene mutation [13,14], provides indirect evidence that a single PIG-A gene mutation is not sufficient for the development of PNH.

In addition, the one-step model fails to address the relationships between PNH and aplastic anemia (AA), which are well described but poorly understood. To develop aplasia, not only would the PIG-A mutant hematopoietic stem cell have an intrinsic growth advantage, but it must also be able to suppress normal hematopoiesis through unknown means. Recent knock-out experiments with PIG-A mutant embryonic stem cells [15,16] and mice [17,18] demonstrated no clear intrinsic growth advantage, however, for cells with a PIG-A gene mutation. Taken together, the data do not support this basic tenet of the one-step model.

Two-step model

The relationship between AA and PNH has been recognized for several decades. Over 30 years ago, Lewis and Dacie noted a positive acidified serum lysis (Ham) test for PNH in 15% of their patients with AA, and also observed that 25% of their patients with PNH developed AA [19]. That same year, Dameshek postulated that PNH and AA arise from a common marrow injury but following different responses [20]. Flow cytometry to enumerate GPI-deficient blood cells has documented that up to 50% of patients with AA have laboratory evidence of PNH [21-23]. In their review [24], Rotoli and Luzzatto discussed hypothetical relationships among PNH, AA, and acute leukemia, and speculated depressed hematopoiesis might be independent of PNH, which could develop anytime during aplasia. Young also separated the development of marrow injury, presumably immune-mediated, from clonal selection leading to PNH [25]. Blending these observations and concepts forms the two-step model (Figure 2).

Fig. 2. The two-step model for the development of PNH. After developing AA, a hematopoietic stem cell acquires a PIG-A mutation, presumably as a random event. Outgrowth of the PIG-A mutant PNH clone only occurs within an injured marrow, hence the PIG-A mutation is necessary but not sufficient for the development of PNH.

The two-step model formally separates the development of AA from the development of PNH. In this model, the PIG-A mutation is necessary but not sufficient for the development of PNH. The etiology of the PIG-A mutation in a hematopoietic stem cell is presumably random, but acquiring the mutation in the setting of an injured marrow provides the opportunity for a cellular growth advantage. For example, if AA is mediated by autoimmune T lymphocytes, then the PIG-A mutant stem cell might avoid immune destruction by lacking GPI-linked surface proteins, enabling it to proliferate relative to its suppressed hematopoietic counterparts [24,25].

Although the two-step model acknowledges the critical relationship between AA and PNH, it does not adress how PIG-A mutations arise in an injured marrow, how PIG-A mutant cells proliferate and expand, why clonal expansion is so highly variable among persons with AA, and the occasional transition to myelodysplasia or leukemia. Like the one-step model, the two-step model is limited in its capacity to explain fully the development of PNH. A revised multistep model of PNH is therefore appropriate.

Proposed Multistep Model of PNH

Both of the two previous models for the development of PNH embody useful concepts and yet are somewhat simplistic. Neither model explains fully the complex clinical and laboratory features that are observed in patients with PNH and related bone marrow failure syndromes. An accurate and useful model must acknowledge, if not elucidate, each of the following issues regarding PNH:

1. Etiology of the PIG-A gene mutations
2. Requirement of marrow injury for the pathogenesis of PNH
3. Selection processes for abnormal PNH hematopoietic cells
4. Clonal expansion of PIG-A mutant, GPI-deficient hematopoietic cells
5. Clonal dominance achieved by PNH cells in some cases
6. Exhaustion of hematopoiesis with evolution to aplasia
7. Occasional transition to myelodysplasia or leukemia

In accordance with this degree of complexity, we propose a multistep model for the development of PNH. Each step in our model involves a distinct genetic process or pathophysiologic event, as illustrated in Figure 3. Individual steps will be discussed subsequently, but an overview of the multistep model follows: the PIG-A gene mutation is initially acquired in a hematopoietic stem cell, but does not gain a growth advantage unless bone marrow injury occurs with diminution of normal hematopoiesis. In this setting of marrow injury, a selection process occurs followed by resistance to apoptosis (perhaps conferred by overexpression of anti-apoptosis genes), allowing surviving cells to expand clonally. These first four steps must occur sequentially for PNH to develop, but additional events can then occur and lead to accelerated cell growth and clonal dominance. Alternatively, the hematopoietic clones may begin to fail and PNH can evolve into aplasia. Finally, if additional chromosomal abnormalities develop, PNH can transition to myelodysplasia and/or leukemia.

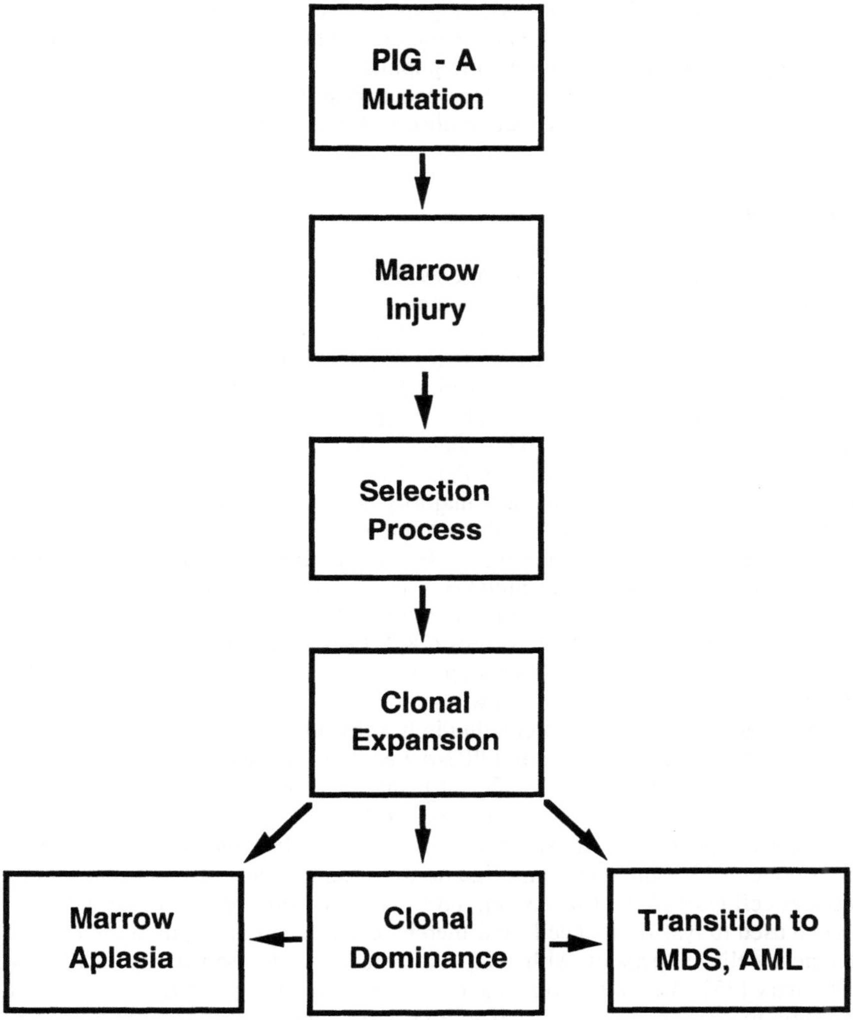

Fig. 3. Proposed multistep model for the development of PNH. Each step includes a distinct process or pathophysiologic event. PIG-A mutations are initially acquired randomly in hematopoietic stem cells, but bone marrow injury or immunosuppression is required to diminish normal hematopoiesis. Selection and preferential survival of PIG-A mutant cells may result from the absence of GPI-linked surface proteins. Additional genetic events, in response to deficient hematopoiesis, lead to resistance to apoptosis and clonal expansion. GPI-deficient cells are now identified by flow cytometry and/or cause clinical symptoms of PNH. In some cases, subsequent events can occur, including dominance of the PNH clone through continued PIG-A mutant cell selection and growth, followed by exhaustion of hematopoiesis to aplasia or transition to myelodysplasia and/or leukemia.

In our proposed multistep model, there are seven distinct steps for the pathogenesis and evolution of PNH (Figure 3). The first four steps must occur sequentially, culminating with sufficient clonal expansion of the PIG-A mutant hematopoietic cells to have measurable populations of GPI-deficient cells in the peripheral blood. The subsequent steps occur only in some persons with PNH, and help explain the clinical variation observed in this disorder.

PIG-A mutation

Since the initial reports of PIG-A gene mutations in patients with PNH, the etiology of these mutations has been a topic of intense interest and investigation. Several important points have been elucidated: (1) PIG-A gene mutations in PNH are always acquired events. To date, germline PIG-A mutations have never been reported; an inherited C55T substitution in the PIG-A gene was proven to be a silent polymorphism [26]; (2) PIG-A gene mutations in PNH appear to be randomly acquired. There is no evidence that the PIG-A gene is inherently hypermutable or that patients with PNH have overall genetic instability. We previously reported that the vast majority of persons with PNH have no increases in the number of acquired DNA mutations in genes outside the PIG-A locus [27]; (3) some persons with PNH have multiple PIG-A mutant clones [13,14]; and (4) PIG-A gene mutations can occur in normal persons. GPI-deficient T lymphocytes emerge rapidly after CAMPATH-1H chemotherapy that lyses CD52-positive (GPI-linked) cells [28,29], an observation that supports the hypothesis that PIG-A mutant hematopoietic cells are present in normal adults. Luzzatto and colleagues used flow cytometry to identify rare GPI-deficient erythroid and myeloid cells circulating in healthy adults, and molecular analyses on pools of cells revealed acquired PIG-A mutations in some cases [30].

The ability to identify GPI-deficient hematopoietic cells improved markedly with the availability of the novel reagent aerolysin, a potent toxin derived from *Aeromonas* species. Aerolysin binds directly to GPI anchors of membrane glycoproteins [31] and forms an oligomeric transmembrane channel that rapidly induces cell lysis [32]. Brodsky and co-workers recently reported that aerolysin can be used to lyse normal cells and thereby enhance small populations of GPI-deficient cells in persons with AA, with much greater sensitivity than flow cytometry [33]. We have developed a novel in vitro aerolysin negative selection strategy to isolate and characterize PIG-A mutant cells [34]. The mutant frequency of aerolysin-resistant T lymphocyte clones in normal adults averaged 18 per million peripheral blood mononuclear cells, very close to the mutant frequency for acquired mutations in the *hprt* gene locus [27]. Aerolysin-resistant T cell clones had a Class A complementation defect and distinct PIG-A gene mutations. Larger numbers of aerolysin-resistant clones were observed in persons recovering from AA compared to those with Fanconi Anemia [34].

To summarize the first step of our proposed multistep model: at least one PIG-A mutation is randomly acquired in hematopoietic progenitor cells, a process that occurs in normal persons and does not automatically portend PNH. The pathogenesis of PNH then involves the intense negative selection and expansion of these naturally occurring PIG-A mutant hematopoietic stem cells.

Marrow injury

As discussed earlier, the PIG-A gene mutation by itself does not appear to be sufficient for the development of PNH. The frequent co-occurrence of PNH with AA, however, suggests that bone marrow aplasia might also be an essential part of the pathogenesis of this disorder. Indeed, virtually all cases of PNH have clinical and laboratory evidence of deficient hematopoiesis [2,23,35]. In many patients there is clinically apparent marrow hypoplasia or even overt AA, while in others the hematopoietic defect is more subtle and characterized by subclinical cytopenia, macrocytosis, and decreased bone marrow cellularity.

Most persons who develop PNH have no exposure to toxins or medications, and have no evidence of viral-induced marrow injury. It is likely, therefore, that autoreactive T lymphocytes mediate the bone marrow suppression, similar to the hematopoietic injury that occurs in persons with AA but without evidence of PNH. Mechanisms of immune-mediated suppression may involve direct contact or soluble factors such as γ-interferon [36]. An altered T cell receptor repertoire in persons with PNH was recently reported [37], consistent with the hypothesis that an immune-mediated process is involved in the pathogenesis of PNH.

To summarize the second step of our proposed multistep model: after the acquisition of one or more PIG-A gene mutations, a distinct pathophysiological process occurs whereby normal hematopoiesis is diminished, probably through autoreactive immune-mediated suppression. If bone marrow injury does not occur, then the PIG-A mutant cells will not have the opportunity for selection and clonal expansion in the setting of diminished hematopoiesis.

Selection process

After suppression of normal hematopoiesis, whether by an immune-mediated process or other injury to the bone marrow, there may be very few remaining hematopoietic stem cells. If the marrow injury is severe, the patient will develop AA and recover hematopoiesis only with aggressive immunosuppressive therapy or bone marrow transplantation. If the bone marrow injury is incomplete, however, then the remaining hematopoietic stem cells, especially naturally occurring PIG-A mutant cells, can be selected for growth and expansion. In a patient with PNH who underwent transplantation with healthy syngeneic stem cells, the abnormal PIG-A mutant cells had a survival advantage relative to the transplanted cells [38]. These data support the hypothesis that PNH includes a process whereby PIG-A mutant cells are selected for growth.

The mechanisms by which surviving PIG-A mutant hematopoietic stem cells might be selected for growth are currrently not well understood. Perhaps autoreactive T lymphocytes mediate suppression of hematopoiesis through direct contact with certain GPI-linked surface proteins [39], thereby allowing the GPI-deficient PNH cells to escape immunosuppression. While this hypothesis is attractive, a recent study failed to demonstrate such selective suppression [40]. The scenario of autoreactive suppression of normal hematopoiesis by GPI-linked surface proteins has not been proven in patients with PNH at this time, although investigation on lymphocytes and stem cell antigens may soon reveal insights.

To summarize the third step of our proposed multistep model: naturally occurring PIG-A mutant hematopoietic stem cells survive bone marrow injury and are preferentially selected for growth to help attain marrow recovery.

Clonal expansion

The fourth required step for the development of PNH is the clonal expansion of selected PIG-A mutant cells. Without this process, PNH clones would remain relatively small and unable to produce sufficient progeny to detect GPI-deficient cells in the peripheral blood. While the PIG-A mutant cells could have an intrinsic growth advantage over remaining hematopoietic cells [41], this is difficult to assess in the setting of an injured marrow with few stem cells.

We previously documented that PNH cells do not have a proliferative advantage over their normal counterparts, and proposed that a survival advantage might confer a cellular growth advantage [42]. Brodsky and colleagues reported that PNH granulocytes had relative resistance to apoptosis [43], and this resistance to apoptosis was also found in other bone marrow failure conditions including AA and MDS [44]. We later confirmed resistance to apoptosis occurred in PNH, but noted that restoration of GPI-linked proteins did not affect the resistant phenotype [45]. Taken together, these data indicate that resistance to apoptosis occurs in PNH and related bone marrow failure conditions; the phenomenon reflects a common survival mechanism for cells expanding from an injured marrow, rather than a process unique to PNH.

Since resistance to apoptosis may help explain clonal expansion of PIG-A mutant cells within an injured bone marrow, we recently investigated the molecular basis for this phenomenon. Using pooled granulocyte RNA from patients with PNH compared to RNA from normal granulocytes, we identified 4 anti-apoptosis genes that are 2-4 times overexpressed in PNH: human A1, hHR23B, Mcl-1, and RhoA [46]. One or more of these genes may help confer the phenotype of resistance to apoptosis, and thereby provide the growth advantage leading to clonal expansion of both GPI-negative and GPI-positive remaining marrow cells. Modest upregulation of anti-apoptosis genes might allow clonal expansion without driving clones toward malignant transformation.

To summarize the fourth step of our proposed multistep model: genetic events in response to marrow injury include upregulation of anti-apoptosis genes. The resulting resistance to apoptosis confers a growth advantage for selected cells, which leads to clonal expansion. Once PNH clones reach >3% population in the peripheral blood, then they are easily detected by flow cytometry [47]. At this point the patient should be characterized as having PNH.

Additional steps in the model

Once a person has developed PNH with measurable populations of GPI-deficient cells in the peripheral blood, the size of the PNH clone may remain stable for many years. In some patients, however, additional processes can occur including clonal dominance, marrow aplasia, and transition to MDS or AML.

Clonal dominance

At least half of the patients with PNH will have marked growth of GPI-deficient cells to a level of clonal dominance, sometimes reaching >90% of the circulating peripheral granulocytes and erythrocytes. This process is especially remarkable because it indicates that a single PIG-A mutant cell (or very few) can support human hematopoiesis.

In our proposed multistep model, clonal dominance represents a distinct process that likely has a molecular basis. Genes involved in this phenomenon have not been elucidated, but could include TAXREB107, AF17, and EGR-1 [48] or the WT1 gene [49]. Upregulation of these genes might accelerate the process of clonal expansion, but also might inhibit normal hematopoiesis and thereby assist the clonal dominance.

Marrow aplasia

As observed over 30 years ago, PNH can evolve into AA. In our proposed model, the clonally expanded (and occasionally clonally dominant) PNH cells simply exhaust their ability to sustain hematopoiesis after a period of months to years. This terminal event in the sequence reflects the difficulty of a limited number of surviving stem cells to remain clonally expanded and support hematopoiesis indefinitely [50].

Transition to MDS or AML

In a small percentage of patients, additional genetic events can change the clinical phenotype from PNH to MDS. Beyond the gene upregulation that leads to clonal expansion and later clonal dominance, generalized chromosomal instability can develop within the PNH clone and lead to an abnormal karyotype [51]. Dysplastic changes are then usually observed in the marrow and the patients has a proper diagnosis of MDS with a clear preleukemic condition.

If even more genetic events occur, such as proto-oncogene upregulation or tumor suppressor gene inactivation, then an overt malignant transformation can develop within the PNH clone. Specific genetic events required for leukemic transformation are unknown in PNH, just as they are unknown for de novo leukemia. One PNH patient who developed AML within his GPI-deficient clone had no clear abnormalities in c-myb, c-myc, c-fos, or c-fms activation [52].

Conclusions

Our proposed multistep model of PNH, compared to the simpler one-step and two-step models, addresses many puzzling aspects of this disorder. Separating the pathogenesis and evolution of PNH into distinct individual steps will encourage focused research on each particular pathophysiologic event, with a goal of elucidating mechanisms and genes involved in each step. This approach should help our understanding of the mysterious disease known as PNH.

References

1. Rosse WF, Parker CJ (1985) Paroxysmal nocturnal haemoglobinuria. Clin Haematol 14:105-125

2. Ware RE, Hall SE, Rosse WF (1991) Paroxysmal nocturnal hemoglobinuria with onset in childhood and adolescence. N Engl J Med 325:991-996

3. Nicholson-Weller A, March JP, Rosenfeld SI, et al (1983) Affected erythrocytes of patients with paroxysmal nocturnal hemoglobinuria are deficient in the complement regulatory protein, decay accelerating factor. Proc Natl Acad Sci USA 80:5066-5070

4. Shichishima T, Terasawa T, Hashimoto C, et al (1991) Heterogeneous expression of decay accelerating factor and CD59/membrane attack complex inhibition factor on paroxysmal nocturnal haemoglobinuria (PNH) erythrocytes. Br J Haematol 78:545-550

5. Takahaski M, Takeda J, Hirose S, et al (1993) Deficient biosynthesis of N-Acetylglucosaminyl-phosphatidyl-inositol, the first intermediate of glycosyl phosphatidyl-inositol anchor biosynthesis, in cell lines established from patients with paroxysmal nocturnal hemoglobinuria. J Exp Med 177:517-521

6. Norris J, Hall S, Ware RE, et al (1994) Glycosyl-phosphatidylinositol anchor synthesis in paroxysmal nocturnal hemoglobinuria: Partial or complete defect in an early step. Blood 83:816-821

7. Miyata T, Takeda J, Iida Y, et al (1993) The cloning of PIG-A, a component in the early step of GPI-anchor biosynthesis. Science 259:1318-1320

8. Ware RE, Rosse WF, Howard TA (1994) Mutations within the Piga gene in patients with paroxysmal nocturnal hemoglobinuria. Blood 83:2418-2422

9. Takeda J, Miyata T, Kawagoe K, et al (1993) Deficiency of the GPI anchor caused by a somatic mutation of the PIG-A gene in paroxysmal nocturnal hemoglobinuria. Cell 73:703-711

10. Ware RE, Rosse WF, Howard TA (1994) Mutations within the Piga gene in patients with paroxysmal nocturnal hemoglobinuria. Blood 83:2418-2422

11. Bessler M, Mason PJ, Hillmen P, et al (1994) Paroxysmal nocturnal hemglobinuria (PNH) is caused by somatic mutations in the PIG-A gene. EMBO J 13:110-117

12. Rosse WF, Ware RE (1995) The molecular basis of paroxysmal nocturnal hemoglobinuria. Blood 86:3277-3286

13. Endo M, Ware RE, Vreeke TM, et al (1996) Molecular basis of the hetergeneity of expression of glycosyl phosphatidylinositol anchored proteins in paroxysmal nocturnal hemoglobinuria. Blood 87:2546-2557

14. Nishimura J, Inoue N, Wada H, et al (1997) A patient with paroxysmal nocturnal hemoglobinuria bearing four independent PIG-A mutant clones. Blood 89:3470-3476

15. Kawagoe K, Kitamura D, Okabe M, et al (1996) Glycosylphosphatidylinositol-anchor-deficient mice: Implications for clonal dominance of mutant cells in paroxysmal nocturnal hemoglobinuria. Blood 87:3600-3606

16. Rosti V, Tremml G, Soares V, et al (1997) Murine embryonic stem cells without pig-a gene activity are competent for hematopoiesis with the PNH phenotype but not for clonal expansion. J Clin Invest 100:1028-1036

17. Tremml G, Dominguez C, Rosti V, et al. (1999) Increased sensitivity to complement and a decreased red blood cell life span in mice mosaic for a nonfunctional *Piga* gene. Blood 94:2945-2954

18. Murakami Y, Kinoshita T, Maeda Y, et al (1999) Different roles of glycosylphosphatidylinositol in various hematopoietic cells as revealed by a mouse model of paroxysmal nocturnal hemoglobinuria. Blood 94:2963-2970

19. Lewis SM, Dacie JV (1967) The aplastic anemia-Paroxysmal nocturnal hemoglobinuria syndrome. Br J Haematol 13:236-251

20. Dameshek W (1967) Riddle: What do aplastic anemia, paroxysmal nocturnal hemoglobinuria (PNH) and "hypoplastic" leukemia have in commone? (editorial) Blood 30:251-254.

21. Schubert J, Vogt HG, Zielinska-Skowronek M, et al (1994) Development of the glycosylphosphatidylinositol-anchoring defect characteristic for paroxysmal nocturnal hemoglobinuria in patients with aplastic anemia. Blood 83:2323-2328

22. Nakakuma H, Nagakura S, Iwamoto N, et al (1995) Paroxysmal nocturnal hemoglobinuria clone in bone marrow patients with pancytopenia. Blood 85:1371-1376

23. Dunn DE, Tanawattanacharoen P, Boccuni P, et al (1999) Paroxysmal nocturnal hemoglobinuria cells in patients with bone marrow failure syndromes. Ann Intern Med 131:401-408

24. Rotoli B, Luzzatto L (1989) Paroxysmal nocturnal haemoglobinuria. Bailliere's Clinical Haematology 2:113-138

25. Young NS (1992) The problem of clonality in aplastic anemia: Dr Dameshek's riddle, restated. Blood 79:1385-1392

26. Endo M, Ware RE, Vreeke TM, Howard TA, Parker CJ (1996) Identification and characterization of an inherited mutation of PIG-A in a patient with paroxysmal nocturnal haemoglobinuria. Br J Haematol 93:590-593

27. Purow DB, Howard TA, Marcus SJ, et al (1999) Genetic instability and the etiology of somatic PIG-A mutations in paroxysmal nocturnal hemoglobinuria. Blood Cells Mol Dis 25:81-91

28. Hertenstein B, Wagner B, Bunjes D, et al. (1995) Emergence of CD52-, phosphatidylinositolglycan-anchor-deficient T lymphocytes after in vivo application of Campath-1H for refractory B-cell non-Hodgkin lymphoma. Blood 86:1487-1492

29. Rawstron AC, Rollinson SJ, Richards S, et al. (1999) The PNH phenotype cells that emerge in most patients after CAMPATH-1H therapy are present prior to treatment. Br J Haematol 107:148-153

30. Araten DJ, Nafa K, Pakdeesuwan K, et al (1999) Clonal populations of hematopoietic cells with paroxysmal nocturnal hemoglobinuria genotype and phenotype are present in normal individuals. Proc Natl Acad Sci USA 96:5209-5214

31. Diep DB, Nelson KL, Raja SM, et al (1998) Glycosylphosphatidylinositol anchors of membrane glycoproteins are binding determinants for the channel-forming toxin aerolysin. J Biol Chem 273:2355-2360

32. Parker MW, Buckley JT, Postma JPM, et al (1994) Structure of the *Aeromonas* toxin proaerolysin in its water-soluble and membrane-channel states. Nature 367:292-295

33. Brodsky RA, Mukhina GL, Nelson KL, et al (1999) Resistance of paroxysmal nocturnal hemoglobinuria cells to the glycosylphosphatidylinositol-binding toxin aerolysin. Blood 93:1749-1756

34. Ware RE, Pickens CV, DeCastro DM, et al (In Press) Circulating PIG-A mutant T lymphocytes in healthy adults and patients with bone marrow failure syndromes. Exp Hematol

35. Maciejewski JP, Sloand EM, Sato T, Anderson S, Young NS (1997) Impaired hematopoiesis in paroxysmal nocturnal hemoglobinuria/aplastic anemia is not associated with a selective proliferative defect in the glycosylphosphatidylinositol-anchored protein-deficient clone. Blood 89:1173-1181

36. Yu JM, Emmons RV, Hanazono Y, et al (1999) Expression of interferon-gamma by stromal cells inhibits murine long-term repopulating hematopoietic stem cell activity. Exp Hem 27:895-903

37. Karadimitris A, Manavalan JS, Thaler HT, et al (2000) Abnormal T-cell repertoire is consistent with immune process underlying the pathogenesis of paroxysmal nocturnal hemoglobinuria. Blood 96:2613-2620

38. Endo M, Beatty PG, Vreeke TM et al (1996) Syngeneic bone marrow transplantation without conditioning in a patient with paroxysmal nocturnal hemoglobinuria: in vivo evidence that the mutant stem cells have a survival advantage. Blood 88:742-750

39. Young NS (2000) Hematopoietic cell destruction by immune mechanisms in acquired aplastic anemia. Semin Hematol 37:3-14

40. Karadimitris A, Notaro R, Koehne G, et al (2000) PNH cells are as sensitive to T-cell-mediated lysis as their normal counterparts: implications for the

pathogenesis of paroxysmal nocturnal hemoglobinuria. Br J Haematol 111:1158-1163

41. Iwamoto N, Kawaguchi T, Horikawa K, et al (1996) Preferential hematopoiesis by paroxysmal nocturnal hemoglobinuria clone engrafted in SCID mice. Blood 87:4944-4948

42. Tseng JE, Hall SE, Howard TA, Ware RE (1995) Phenotypic and functional analysis of lymphocytes in paroxysmal nocturnal hemoglobinuria. Am J Hematol 50:244-253

43. Brodsky RA, Vala MS, Barber JP, et al (1997) Resistance to apoptosis caused by *PIG-A* gene mutations in paroxysmal nocturnal hemoglobinuria. Proc Natl Acad Sci USA 94:8756-8760

44. Horikawa K, Nakakuma H, Kawaguchi T, et al. (1997) Apoptosis resistance of blood cells from patients with paroxysmal nocturnal hemoglobinuria, aplastic anemia, and myelodysplastic syndrome. Blood 90:2716-2722

45. Ware RE, Nishimura J, Moody MA, et al (1998) The PIG-A mutation and absence of glycosylphosphatidylinositol-linked proteins do not confer resistance to apoptosis in paroxysmal nocturnal hemoglobinuria. Blood 92:2541-2550

46. Ware RE, Ormsbee SM, Moody MA, et al (2000) Overexpression of anti-apoptosis genes in association with resistance to apoptosis observed in bone marrow failure syndromes. Blood 96:526a (abstract)

47. Hall SE, Rosse WF (1996) The use of monoclonal antibodies and flow cytometry in the diagnosis of paroxysmal nocturnal hemoglobinuria. Blood 87:5332-5340

48. Schubert J, Liakicheva A, Felda O, et al (2000) Differential gene expression in GPI-deficient neutrophils from patients with paroxysmal nocturnal hemoglobinuria. Blood 96:232a (abstract)

49. Ikeda K, Shichishima T, Okamoto M, et al (2001) High expression of the Wilms tumor WT1 gene RNA in patients with paroxysmal nocturnal hemoglobinuria (PNH). Blood (abstract)

50. Maciejewski JP, Kis S, Sloand E, et al (2000) Sustained long-term hematologic recovery despite a marked quantitative defect in the stem cell compartment of patients with aplastic anemia after immunosuppressive therapy. Am J Hematol 65:123-131

51. Nishimura J, Inoue N, Azenishi Y, et al (1996) Analysis of *PIG-A* gene in a patient who developed reciprocal translocation of chromosome 12 and paroxysmal nocturnal hemoglobinuria during follow-up of aplastic anemia. Am J Hematol 51:229-233

52. Stafford HA, Nagarajan S, Weinberg JB, et al (1995) PIG-A, DAF, and proto-oncogene expression in paroxysmal nocturnal hemoglobinuria-associated acute myelogenous leukemia blasts. Br J Haematol 89:72-78

Proposals for Classification of the Clinical Stages, Grading of Severity and the Molecular Pathogenesis of Paroxysmal Nocturnal Hemoglobinuria

Tsutomu Shichishima

First Department of Internal Medicine, Fukushima Medical University, 1 Hikariga-oka, Fukushima, Fukushima 960-1295, Japan

Summary: Paroxysmal nocturnal hemoglobinuria (PNH) is an acquired disorder caused by complement-mediated hemolysis of complement-sensitive erythrocytes and bone marrow failure. A glycosylphosphatidylinositol glycan-class A (PIG-A) gene abnormality at the level of hematopoietic precursors is essential for the occurrence of PNH and explains the deficiency of glycosylphosphatidylinositol (GPI)-anchored membrane proteins in various hematopoietic cells from PNH patients which elicits complement-mediated hemolysis. Quantitation of affected cells of various hematopoietic cell lineages by flow cytometry using monoclonal antibodies to GPI-anchored membrane proteins enables classification of the clinical stages and grading of severity of PNH. However, it is not known how an abnormality of the PIG-A gene occurs or how a PNH clone with a PIG-A gene mutation proliferates or is selected in bone marrow, although some hypotheses have been proposed. We found a high frequency of the HLA-DRB1*1501-DQA1*0102-DQB1*0602 haplotype and high expression of the Wilms' tumor gene, WT1, in Japanese primary PNH and aplastic anemia-PNH syndrome patients. These facts suggest that an immune mechanism in an HLA-restricted manner plays an important role in the occurrence of a PNH clone, and that a PNH clone increases or decreases relatively or absolutely through competition between proliferation of WT1-expressing cells and clonal selection against WT1-expressing cells.

Key words: Paroxysmal nocturnal hemoglobinuria (PNH), Classification of clinical stage, Classification of grading of severity, HLA class II haplotype, Wilms' tumor gene (WT1)

Clinical Features and Bone Marrow Failure

Paroxysmal nocturnal hemoglobinuria (PNH) is an acquired hemolytic disorder caused by complement-mediated hemolysis of complement-sensitive erythrocytes [1]. PNH erythrocytes consist of complement-sensitive (PNH II and III) and –

insensitive (PNH I) erythrocytes based on differences in their sensitivity to complement according to the complement lysis sensitivity (CLS) test [2]. Hypersensitivity to complement as judged by the CLS test is also found in granulocytes, platelets, and hematopoietic precursor cells [3-5]. Thus, PNH is considered to be a stem cell disorder affecting all hematopoietic lineages [1]. The affected hematopoietic cells lack glycosylphosphatidylinositol (GPI)-anchored membrane proteins on their surfaces due to a phosphatidylinosistol glycan-class A (PIG-A) gene mutation [6]. The differences in the sensitivity of PNH erythrocytes to complement is related to the extent of expression of CD55 (decay-accelerating factor) and CD59, complement regulatory proteins, and the phenotypes of PNH erythrocytes determined by the CLS test are similar to those determined by flow cytometry [7, 8]. Thus, flow cytometric single-color or two-color analysis using CD55 and/or CD59 monoclonal antibodies is utilized to diagnose PNH and to determine the phenotypes of PNH erythrocytes. The flow cytometric method is superior, with respect to both sensitivity and specificity for detecting PNH erythrocytes, to the old methods which employ complement-mediated hemolysis in vitro, i. e. Ham's test, sugar-water test, and the CLS test [8, 9]. Recently, having over 1% CD59-negative cells was considered positive for PNH erythrocytes [9]. The proportion of circulating GPI-deficient reticulocytes in PNH patients is typically higher than that of affected erythrocytes [10]. More recently, a new diagnostic test for PNH was developed which is based on the ability of a fluorescent labeled inactive variant of the protein aerolysin (FLAER) to bind selectively to GPI anchors [11]. CD55 [12], a 70 kilodalton (KD) membrane protein involved in the inhibition of the complement cascade at the level of C3 convertase and C5 convertase, and CD59 [13], an 18 KD membrane protein which also inhibits the formation of a membrane attack complex of complement by inhibiting the binding of C9 to C5b-8 and of C9 polymerization, are deficient on complement-sensitive erythrocytes [7, 8, 14]. Rare individuals with the Inab phenotype whose erythrocytes lack the Cromer-related blood group antigens and have an inherited deficiency of CD55 alone have been described. Although the sensitivity to complement-mediated lysis of the Inab phenotype erythrocytes was slightly higher than that of normal erythrocytes [15], there have not been any reports of in vivo hemolysis in these individuals [15]. On the other hand, a case of PNH with deficiency of only CD59 on erythrocytes was reported and designated as inherited complete homologous restriction factor (HRF)-20 deficiency by Yamashina et al [16]. The sensitivity of erythrocytes to complement-mediated lysis by the CLS test in this patient was higher than that of normal and Inab phenotype erythrocytes, but lower than that of PNH III erythrocytes, probably due to the presence of CD55 on the cell surfaces [17]. This fact suggests that PNH III erythrocytes determined by the CLS test lack both CD55 and CD59, but not CD55 or CD59 alone, and are characteristic in PNH. Clinically, complement-mediated hemolysis of complement-sensitive erythrocytes in PNH patients leads to hemolytic anemia, hemoglobinuria or hemosiderinuria, hemolytic icterus and sometimes complications of iron deficiency anemia, acute renal failure following hemolytic crisis, aplastic or megaloblastic crisis, cholelithiasis, and venous or arterial thrombosis in patients with PNH. Most of the hematologic, laboratory, biochemical and urinalysis findings in PNH patients are explained by chronic or acute intravascular hemolysis. Although many clinical features of PNH are caused by complement-mediated hemolysis of complement-sensitive erythrocytes, cytopenias, including leucocytopenia, thrombocytopenia, and pancytopenia, and hypocellular bone marrow can not be explained by this mechanism but they are also important features of PNH. The cytopenias seem to be due generally to hypocellular bone marrow. In fact, analysis of the life span in vivo of platelets and granulocytes showed no differences between PNH patients and healthy volunteers [18], although hypersensitivity to complement in vitro of platelets and granulocytes from PNH patients compared to those from healthy volunteers has been demonstrated [3]. It has been reported that 32% or 40.4% of patients with PNH have pancytopenia or hypocellular bone marrow, respectively [19, 20]. The phenomenon of aplastic anemia (AA) patients acquiring PNH is well known as the AA-PNH syndrome described by Lewis and Dacie [21]. Several recent reports [19, 20, 22, 23] indicated that PNH eventually develops in 28.8% to 45% of AA patients treated with immunosuppressive agents and/or androgens. In addition, typical PNH is diagnosed and pancytopenia develops later. Although this was thought to be uncommon, it occurs in 15% of PNH patients [19]. Also, it has been reported that myelodysplastic syndrome (MDS) developed in nine of 220 patients with PNH in a long-term follow-up [19]. In contrast, Dunn et al [24] reported GPI-anchored protein-deficient cells in nine of 39 (23%) patients with MDS. Thus, it is considered that PNH, AA, and MDS arise from a common abnormal property of bone marrow, designated as bone marrow failure syndromes or hypoplastic syndromes [25]. However, it is mostly unknown why hypocellular bone marrow occurs in patients with PNH. Clinically, cytopenias due to hypocellular bone marrow cause the complications of hemorrhagic diathesis and infection in patients with PNH, and these are the main causes of death of Japanese PNH patients [20]. It has been reported that 0~4.8% of PNH patients develop acute leukemia [26, 27]. Most patients with acute leukemia arising from PNH are diagnosed as having acute non-lymphocytic leukemia, including acute myelogenous leukemia, acute myelomonocytic leukemia, and erythroleulemia. With respect to the timing of the development of leukemia/dysplasia once PNH is present, intervals ranging from weeks or a few months to 25 years have been reported [28]. Leukemia/dysplasia blasts originate from PNH clones with deficiencies in GPI-anchored protein expression [26, 27, 29], but in some cases, they arise from GPI-positive cells [30]. It is unclear why acute leukemias arise from PNH in spite of some hypotheses and experiments, involving the ras and RB1 genes [28]. In these cases, the clinical and laboratory manifestations of PNH disappear or become normal with the onset of leukemia [27]. The prognosis of acute leukemia arising from PNH is very poor.

Table 1. Clinical stages in PNH

Findings	Early PNH	Overt PNH	Late PNH
GPI protein deficiency			
Erythrocytes	(-)	(+)	(+ ~ -)
Granulocytes, Monocytes, and Platelets	(+)	(+)	(+ ~ -)
T lymphocytes	(-)	(- ~ +)	(+ ~ -)
Hemolysis	(-)	(+)	(- ~ +)
Marrow hypoplasia	(±)	(±)	(± ~ +)

Proposals for Classification of the Clinical Stages and Grading of Severity of PNH

Twenty kinds of GPI-anchored membrane proteins are deficient in various hematopoietic cells (erythrocytes, neutrophils, monocytes, platelets, T and B cells, natural killer cells, and hematopoietic precursor cells) from patients with PNH, as summarized by Boccuni et al [31]. It has been reported that the deficiency of GPI-anchored proteins generally appears first in granulocytes, monocytes, and platelets, second in erythrocytes, and lastly in lymphocytes from the peripheral blood of PNH patients [32]. This allows for the classification of clinical stages in PNH into early, overt and late PNH (Table 1), as Nakakuma et al [33] also proposed.

Early PNH is defined as a stage with deficiencies of GPI-anchored proteins on granulocytes, monocytes, and platelets, but not on erythrocytes. In this stage, there are no clinical, hematologic and urinalysis findings related to intravascular hemolysis, but there are principally those related to hypocellular bone marrow. Overt PNH is defined as a stage with deficiencies of the GPI-anchored proteins, CD55 and CD59, on erythrocytes. In this stage, to varying degrees there are clinical, laboratory, hematologic and urinalysis findings associated with intravascular hemolysis are found much to less. The bone marrow varies from hypercellular to hypocellular. Late PNH is defined as a stage transforming from the pathophysiology of overt PNH to that of AA, in which the bone marrow is hypocellular and none or little complement-sensitive erythrocytes are found, but deficiencies of GPI-anchored proteins on lymphocytes continue if they appeared in the overt stage of PNH. However, further studies are needed to establish the border between overt and late PNH. It was reported that it took 2.5 years to evolve from AA to overt PNH and 5 years from AA to MDS in treated AA patients [34]. Therefore, we should examine by flow cytometry various hematopoietic cells using monoclonal antibodies to GPI-anchored proteins, in addition to hematologic and clinical examinations, to know the clinical stages during a long-term clinical course in patients with PNH.

Grading the severity of PNH would be useful for planning the treatment of patients with PNH. The proportion of complement-sensitive erythrocytes partly indicates the clinical severity as does the frequency of exacerbation of hemoglobinuria, as Rosse [35] and Shichishima et al [36] reported, respectively. The clinical features found in patients with PNH are determined principally by bone marrow cellularity in addition to the proportion of complement-sensitive erythrocytes and in some cases, by the complications [19, 20, 23]. It is possible to determine accurately the proportion of complement-sensitive erythrocytes by flow cytometry using monoclonal antibodies to CD55 and CD59 [8, 9]. This enables us to grade the severity of PNH, which can be classified into mild, moderate, and serious PNH (Table 2).

Table2. Grading the severity of PNH

Items	Mild	Moderate	Serious
A. Proportion of C'-sensitive erythrocytes	0 1~30%	1 30~60%	2 ≧60%
Frequency of exacerbated hemoglobinuria or hemolytic crisis	0 No hemolytic attacks or one attack every over 3 months	1 One attack every 1 to 3 months	2 More than one attack per month
B. Bone marrow aspirate	0 Normocellular or Hypercellular	1 Hypocellular (3~10 x10^4 cells/μl)	2 Severe hypocellular (<3 x10^4 cells/μl)
C. Thrombosis	0 None	1 Superficial-vein thrombosis	2 Deep-vein thrombosis (or Arterial thrombosis)
D. Severe complications associated with PNH	0 None	1 Observed	-

Grading the severity of PNH is usually determined by a total score of items A and B: "Mild", 0; "Moderate", 1 to 3; and "Serious", 4. The grading is determined by a total score of items C and D, independent of the above criteria, when PNH patients develop thrombosis and severe complications, including severe infection, renal failure, hemosiderosis and acute leukemia. The complications should be properly treated in patients with a score >1.
C' means complement.

Only patients with mild PNH need to be watched carefully. Patients with moderate or serious PNH need to be treated with prednisolone [37], androgens [37, 38], antithymocyte globulin (ATG), etc. Hematopoietic stem cell transplantation should be tried when PNH patients have severe hypocellular bone marrow, repeated hemolytic crises, and repeated arterial- and/or veinous-thromboses in association with a poor prognosis. In addition, some patients with serious PNH should be considered for hematopoietic stem cell transplantation even in an early stage of PNH because it can be difficult to achieve an effective therapy with conventional drugs reported in the past [39].

Backgroud on Proliferation or Selection of a PNH Clone

All the PNH patients reported in the past who were examined for PIG-A gene abnormalities expressed abnormal mRNA and/or genomic DNA in granulocytes, mononuclear cells, myeloid and erythroid progenitor cells, leukemic cells transformed from PNH cells, or in some affected cell lines established from PNH cells [28, 40]. There is an abnormality in the first step of GPI-anchor biosynthesis, the transfer of N-acetylglucosamine to phosphatidylinositol, in affected T- and B-lymphocytic cell lines and granulocytes from PNH patients [41]. This first step in GPI-anchor biosynthesis is regulated by at least four different genes for class A, C and H and GPI1 [42]. Recently, PIG-P was also reported to be associated with this step [43]. Complementation analysis by cell hybridization indicates that affected T- and B-cell lines established from PNH patients have a class A abnormality. cDNA was cloned and complemented CD55 and CD59 expression on the surface membranes of class A-deficient mutants [44]. The PIG-A gene is located at Xp 22.1 on the short arm of the X chromosome [45]. The PIG-A gene cDNA contains 3589 base pairs and encodes a protein of 484 amino acids [46]. PIG-A gene mutations have been found in all the PNH patients reported in the past who have been examined for PIG-A gene abnormalities (Table 3).

Table 3-A Differences in the frequency of the places mutated among various mutations of the PIG-A gene

Mutations	Exon(%)					Intron(%)				Others	Total
	2	3	4	5	6	2	3	4	5	(%)	(%)
Missense	19.0	2.0	0.5	1.5	2.0	-	-	-	-	-	25.0
Nonsense	6.0	1.0	0.5	1.0	3.0	-	-	-	-	-	11.5
Alternative splicing	-	-	0.5	-	-	0.5	2.0	2.5	2.5	-	8.0
Frameshift	26.0	5.5	4.0	7.0	7.0	-	-	0.5	-	-	50.0
Inframe deletion	1.5	-	-	-	-	-	-	-	-	-	1.5
Others	-	0.5	-	-	-	-	-	0.5	1.5	1.5	4.0
Total	52.5	9.0	5.5	8.5	12.0	0.5	2.0	3.5	4.0	1.5	100

Table 3-B Differences in the frequency of the places mutated in the PIG-A gene among various pathophysiologies

Patho-physiolo-gies	Exon(%)					Intron(%)				Others (%)	Total (%)
	2	3	4	5	6	2	3	4	5		
PNH	36.5	5.5	3.0	7.0	9.0	0.5	1.5	2.5	3.5	1.0	70.0
AA-associated	12.5	3.5	2.5	2.5	2.5	-	0.5	1.0	0.5	0.5	26.0
Others	3.5	-	-	-	0.5	-	-	-	-	-	4.0
Total	52.5	9.0	5.5	9.5	12.0	0.5	2.0	3.5	4.0	1.5	100

All the PIG-A gene abnormalities reported in the past include 120 patients with primary PNH, 42 with AA-associated PNH and 6 others, including lymphoma-associated patients and healthy individuals, are shown in Tables 3-A and 3-B. The total number of PIG-A gene abnormalities is 200, the number of different PIG-A gene abnormalities is 168, and the number of identical PIG-A gene abnormalities is 18.

This indicates that somatic mutations of the PIG-A gene appear to be essential for the pathogenesis of PNH. It is well known that PIG-A mutations in PNH hematopoietic cells have no hot spots and most abnormalities of the PIG-A gene in PNH are frameshift mutations, which decrease transferase activity in the first step in GPI-anchor biosynthesis resulting in cells completely deficient in GPI-anchored proteins as well as nonsense mutations [47]. In contrast, several abnormalities of the PIG-A gene in PNH are missense mutations, which partially decrease transferase activity resulting in cells partially deficient in GPI-anchored proteins [47]. However, Noji et al [48] recently showed that PIG-A gene abnormalities determine the phenotypes of granulocytes, but not erythroblasts. In fact, it was found that the amount of GPI-anchored proteins released in exosomes was highly modified when reticulocytes from PNH patients were matured in vitro [49]. Also, transfer of GPI-anchored proteins to endothelium cells from erythrocytes in a transgenic system [50] and to erythrocytes from high-density lipoproteins [51] could account for the differences in phenotype between mature erythrocytes and erythroid precursors. These facts suggest that the PNH phenotypes determined by the expression of GPI proteins can change and be modified during differentiation and maturation of precursors to mature cells. In addition, Noji et al [48] found that the distribution of predominant PNH clones with PIG-A gene abnormalities is often not the same in different cell types, suggesting that a clonal selection or proliferation of PNH clones with PIG-A gene abnormalities occurs independently in various cell lineages during differentiation and maturation. This explains the heterogeneous expression of GPI-anchored membrane proteins among various hematopoietic cell lineages. The significance of PIG-A gene abnormalities in PNH is clarified by showing that a cell population having a PIG-A gene abnormality is a true PNH clone. Currently, approximately 14% of PNH patients analyzed for the PIG-A gene have oligoclonal PIG-A mutations [40].

It has been suggested that proliferation and selection of a PNH clone with a PIG-A gene abnormality is associated with an immunological mechanism involving cytotoxic-T lymphocytes [25], autosomal proliferation of a PNH clone, and resistance to apoptosis of PNH blood cells [52]. However, in spite of many studies the actural mechanism of proliferation and selection of a PNH clone is unclear and controversial. Some reports [53] support the concept that loss of GPI-anchored proteins results in less efficient killing of GPI-deficient hematopoietic target cells by putative autoimmune T-cell clones, causing proliferation or selection of a PNH clone. In addition, abnormalities of T cells derived from PNH patients have been reported [54]. Nevertheless, a wide variety of immunologic assays did not successfully show a difference between normal and PIG-A mutated target cells [55]. In chimeric knockout mice, the cells without a PIG-A gene constitute a minor, static fraction of bone marrow or circulating hematopoietic cells, and PIG-A minus and normal cells behave similarly in paired tissue culture assays, denying autosomal proliferation of a PNH clone [56, 57]. Moreover, although resistance to apoptosis of granulocytes and $CD34^+$ cells from PNH patients has been demonstrated [52], both GPI-deficient hematopoietic cells with a PIG-A gene abnormality and GPI-positive cells have this property, theoretically suggesting that apoptosis resistance does not result in the proliferation or selection of only a PNH clone. In fact, there is no known molecular mechanism whereby GPI-anchored proteins influence apoptosis, although there are conflicting data [58].

Molecular Pathogenesis in PNH

It is considered that PNH, AA, and MDS result from a common abnormality in bone marrow. At present, although it is certain that hypoplastic bone marrow and decreased proliferative capacity of their precursor cells are common and important features of the three disorders [25], it is unclear what other common characteristics they may have. A PIG-A gene abnormality at the level of hematopoietic precursor cells is essential for the occurrence of PNH and it explains the deficiency of GPI-anchored membrane proteins in various hematopoietic cells from PNH patients and the complement-mediated hemolysis of complement-sensitive erythrocytes in PNH patients. However, it is not known why an abnormality of the PIG-A gene occurs in PNH hematopoietic precursors. It is possible that a PIG-A gene abnormality could result from bone marrow failure, especially hypocellular bone marrow, because this circimstance could lead to gene instability in bone marrow hematopoietic precursor cells. This hypothesis was substantiated by analyses of the phenotypes of glycophorin A on erythrocytes from PNH patients and of microsattelite genes on granulocytes from the patients [59, 60]. Recently, we found that the HLA-DRB1*1501-DQA1*0102-DQB1*0602 haplotype, which determines the presentation of HLA-DR2 in Japanese, is a characteristic of Japanese patients with primary PNH and AA-PNH syndrome compared to healthy Japanese individuals (our unpublished data). This strongly suggests that immune mechanisms in an HLA-restricted manner play an important role in the molecular pathogenesis of primary PNH and AA-PNH syndrome, especially including relationship between occurrence of a PIG-A gene mutation and immune mechanisms in an HLA-restricted manner. It may also be possible that the specific haplotype or other genes linked to it causes the circumstances of gene instability in bone marrow hematopoietic precursor cells by somewhat mechanisms [60]. Accordingly, it is of great interest to study whether the healthy population with a PIG-A gene abnormality is consistent with the healthy population with the specific haplotype because this result would clarify a background of occurrence of a PNH clone. However, it is unclear whether the specific haplotype or other genes linked to it are related to the proliferation of a PNH clone. Finally, there is the problem of why bone marrow failure or hypoplastic bone marrow occurs in PNH patients. Recently, we found higher expression of the Wilms' tumor gene WT1 RNA in bone marrow cells from AA, PNH, and MDS patients compared to healthy volunteers (our unpublished data). It should be considered whether the WT1 gene acts as a tumor suppressor gene or an oncogenic gene is correlated with the differences in interactions of the WT1 gene protein with other regulatory molecules [61]. Recently, some reports indicated that the WT1 gene product is an attractive tumor antigen capable of eliciting cytotoxic-T lymphocytes or helper/inducer-T lymphocytes against WT1-expressing tumors in an HLA-restricted manner [62, 63]. These facts suggest that high expression of the WT1 gene in bone marrow precursor cells from PNH patients may also elicit cytotoxic-T lymphocytes or helper/inducer-T lymphocytes which eliminate the WT1-expressing cells, suggesting that the expression of WT1 RNA may lead to clonal selection of bone marrow precursor cells. If high expression of WT1 RNA in hematopoietic precursor cells from healthy individuals would occur due to acute or chronic viral infection or other factors, the immune mechanism described above may lead to hypocellular bone marrow. In contrast, if high expression of the WT1 gene, acting as an oncogene, occurs in bone marrow precursor cells, it may cause proliferation of a PNH clone as well as MDS and leukemia [61]. The hypothesis proposed from our recent study, based on the dual pathogenesis concept developed by Luzzatto and Bessler [64] and the clonal selection theory developed by Young [25], is summarized in Fig. 1. Unfortunately, we do not have direct evidence about the relatioship between PNH clones and WT1 gene expression and the role of the WT1 gene in bone marrow cells from PNH patients. In the future, these problems should be resolved by clarifying the interactions of the WT1 gene protein with other regulatory proteins and the distribution of WT1 RNA in purified bone marrow precursor cells from PNH patients. However, at present, it is at least possible that gene abnormality other than PIG-A gene mutation in bone marrow cells from PNH patients can occur which may affect proliferation or selection of a PNH clone, suggesting a multi-step mechanism through the background of gene instability in bone marrow precursor cells during the long-term clinical course of PNH.

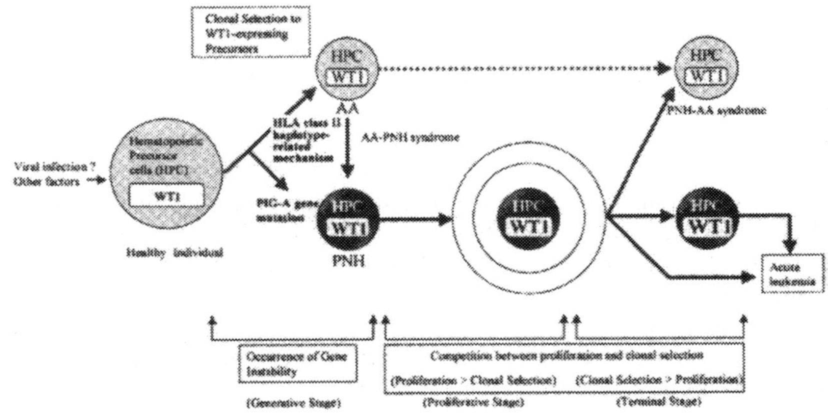

Fig. 1. Hypothesis about the molecular pathogenesis of PNH. The pathogenetic stages of PNH during its clinical course may be divided into three: I: Generative stage, in which a PNH clone with PIG-A gene abnormality occurs due to an immune mechanism in an HLA-restricted manner and gene instability, which may result from clonal selection against WT1-expressing precursors, causing hypocellular bone marrow. II: Proliferative stage, in which the capacity of a PNH clone to proliferate is superior to that of clonal selection against WT1-expressing precursors through competition between proliferation and clonal selection. However, at present, we can not substantiate any proliferation of a PNH clone. III: Terminal stage, in which if an immune mechanism, probably clonal selection against WT1-expressing precursors, succeeds in eliminating them, PNH transforms to AA or there is complete remission of PNH, but if the immune mechanism fails to eliminate them, PNH transforms to MDS or acute leukemia.

References

1. Rosse WF, Parker CJ (1985) Paroxysmal nocturnal haemoglobinuria. Clin Haematol 14: 105-125
2. Rosse WF, Dacie JV (1966) Immune lysis of normal human and paroxysmal nocturnal hemoglobinuria (PNH) red blood cells. I. The sensitivity of PNH red cells to lysis by complement and specific antibody. J Clin Invest 45: 736-748
3. Aster RH, Enright SE (1969) A platelet and granulocyte membrane defect in paroxysmal nocturnal hemoglobinuria: usefulness for the detection of platelet antibodies. J Clin Invest 48: 1199-1210
4. Dessypris EN, Clark DA, McKee LC, et al (1983) Increased sensitivity to complement of erythroid and myeloid progenitors in paroxysmal nocturnal hemoglobinuria. N Engl J Med 309: 690-693

5. Shichishima T, Terasawa T, Uchida T, et al (1989) Complement sensitivity of erythroblasts and erythropoietic precursors in paroxysmal nocturnal haemoglobinuria (PNH). Br J Haematol 72: 578-583
6. Rosse WF, Ware RE (1995) The molecular basis of paroxysmal nocturnal hemoglobinuria. Blood 86: 3277-3286
7. Shichishima T, Terasawa T, Hashimoto C, et al (1991) Heterogeneous expression of decay accelerating factor and CD59/membrane attack complex inhibition factor on paroxysmal nocturnal haemoglobinuria (PNH) erythrocytes. Br J Haematol 78: 545-550
8. Shichishima T, Terasawa T, Saitoh Y, et al (1993) Diagnosis of paroxysmal nocturnal haemoglobinuria by phenotypic analysis of erythrocytes using two-colour flow cytometry with monoclonal antibodies to DAF and CD59/MACIF. Br J Haematol 85: 378-386
9. Hall SE, Rosse WF (1996) The use of monoclonal antibodies and flow cytometry in the diagnosis of paroxysmal nocturnal hemoglobinuria. Blood 87: 5332-5340
10. Iwamoto N, Kawaguchi T, Nagakura S, et al (1995) Markedly high population of affected reticulocytes negative for decay accelerating factor and CD59 in paroxysmal nocturnal hemoglobinuria. Blood 85: 2228-2232
11. Brodsky RA, Mukhina GL, Li S, et al (2000) Improved detection and characterization of paroxysmal nocturnal hemoglobinuria using fluorescent aerolysin. Am J Clin Pathol 114: 459-466
12. Nicholson-Weller A, Burge J, Fearon DT, et al (1982) Isolation of a human erythrocyte membrane glycoprotein with decay-accelerating activity for C3 convertases of complement system. J Immunol 129: 184-189
13. Sugita Y, Mazda T, Tomita M (1989) Amino-terminal amino acids sequence and chemical and functional properties of a membrane attack complex-inhibitory factor from human erythrocyte membranes. J Biochem 106: 589-592
14. Nicholson-Weller A, March JP, Rosenfeld SI, et al (1983) Affected erythrocytes of patients with paroxysmal nocturnal hemoglobinuria are deficient in the complement regulatory protein, decay accelerating factor. Proc Natl Acad Sci USA 80: 5066-5070
15. Merry AH, Rawlinson VI, Uchikawa M, et al (1989) Studies on the sensitivity to complement-mediated lysis of erythrocytes (Inab phenotype) with a deficiency of DAF (decay accelerating factor). Br J Haematol 73: 248-253
16. Yamashina M, Ueda E, Kinoshita T, et al (1990) Inherited complete deficiency of 20-kilodalton homologous restriction factor (CD59) as a cause of paroxysmal nocturnal hemoglobinuria. N Engl J Med 323: 1184-1189
17. Shichishima T, Saitoh Y, Terasawa T, et al (1999) Complement sensitivity of erythrocytes in a patient with inherited complete deficiency of CD59 or with the Inab phenotype. Br J Haematol 104: 303-306
18. Devine DV, Siegel RS, Rosse WF (1987) Interactions of the platelets in paroxysmal nocturnal hemoglobinuria with complement. Relationship to defects in the regulation of complement and of platelet survival in vivo. J Clin Invest 79: 131-137
19. Socie G, Mary J, de Gramont A, et al (1996) Paroxysmal nocturnal haemoglobinuria: long-term follow-up and prognostic factors. Lancet 348: 573-577
20. Fujioka S, Asai T (1989) Prognostic features of paroxysmal nocturnal hemoglobinuria in Japan. Acta Haematol Jpn 52: 1386-1394
21. Lewis SM, Dacie JV (1967) The aplastic anaemia-paroxysmal nocturnal haemoglobinuria syndrome. Br J Haematol 13: 236-251

22. Tichelli A, Gratwohl A, Wursch A, et al (1988) Late haematological complications in severe aplastic anaemia. Br J Haematol 69: 413-418
23. Hillmen P, Lewis SM, Bessler M, et al (1995) Natural history of paroxysmal nocturnal hemoglobinuria. N Engl J Med 333: 1253-1258
24. Dunn DE, Tanawattanacharoen P, Boccuni P, et al (1999) Paroxysmal nocturnal hemoglobinuria cells in patients with bone marrow failure syndromes. Ann Intern Med 131: 401-408
25. Young NS (1992) The problem of clonality in aplastic anemia: Dr Dameshek' riddle, restated. Blood 79: 1385-1392
26. Devine DV, Gluck LL, Rosse WF, et al (1987) Acute myeloblastic leukemia in paroxysmal nocturnal hemoglobinuria. Evidence of evolution from the abnormal paroxysmal nocturnal hemoglobinuria clone. J Clin Invest 79: 314-317
27. Shichishima T, Terasawa T, Hashimoto C, et al (1993) Discordant and heterogeneous expression of GPI-anchored membrane proteins on leukemic cells in a patient with paroxysmal nocturnal hemoglobinuria. Blood 81: 1855-1862
28. Harris JW, Koscick R, Lazarus HM, et al (1999) Leukemia arising out of paroxysmal nocturnal hemoglobinuria. Leukemia and Lymphoma 32: 401-426
29. Stafford AA, Nagarajan S, Weinburg B, et al (1995) PIG-A, DAF and protooncogene expression in paroxysmal nocturnal haemoglobinuria-associated acute myelogenous leukaemia blasts. Br J Haematol 89: 72-78
30. van Kamp H, Smit JW, van den Berg E, et al (1994) Myelodysplasia following paroxysmal nocturnal haemoglobinuria: evidence for the emergence of a separate clone. Br J haematol 87: 399-400
31. Boccuni P, Vecchio LD, Noto RB, et al (2000) Glycosyl phosphatidylinositol (GPI)-anchored molecules and the pathogenesis of paroxysmal nocturnal hemoglobinuria. Oncol Hematol 33: 25-43
32. Nakakuma H, Nagakura S, Iwamoto N, et al (1995) paroxysmal nocturnal hemoglobinuria clone in bone marrow of patients with pancytopenia. Blood 85: 1371-1376
33. Nakakuma H, Kawaguchi T, Horikawa K, et al (1995) Proposal for a clinical classification of the clinical stages of paroxysmal nocturnal hemoglobinuria. Blood 86: 2051-2052
34. Tichilli A, Grawohl A, Nissen C, et al (1994) Late clonal complications in severe aplastic anemia. Leuk Lymphoma 12: 167-175
35. Rosse WF (1980) Paroxysmal nocturnal hemoglobinuria-present status and future prospects. West J Med 132: 219-228
36. Shichishima T, Terasawa T, Uchida T, et al (1991) Relationship between the proportion of paroxysmal nocturnal hemoglobinuria (PNH) III-erythrocytes and the frequency of exacerbated hemoglobinuria in PNH. Eur J Haematol 47: 235-236
37. Rosse WF (1982) Treatment of paroxysmal nocturnal hemoglobinuria. Blood 60: 20-23
38. Paquette RL, Yoshimura R, Veisch C, et al (1997) Clinical characteristics predict response to antithymocyte globulin in paroxysmal nocturnal haemoglobinuria. Br J Haematol 96: 92-97
39. Storb R, Evans RS, Thomas ED, et al (1973) Paroxysmal nocturnal haemoglobinuria and refractory marrow failure treated by marrow transplantation. Br J haematol 24: 743-750

40. Nishimura J, Smith CA, Phillips KL, et al (1998) Paroxysmal nocturnal hemoglobinuria: Molecular pathogenesis and molecular therapeutic approaches. Hemathopathol Mol Hematol 11: 119-146
41. Norris J, Hall S, Ware RE, et al (1994) Glycosyl-phosphatidylinositol anchor synthesis in paroxysmal nocturnal hemoglobinuria: partial or complete defect in an early step. Blood 83: 816-821
42. Watanabe R, Inoue N, Westfall B, et al (1998) The first step of glycosylphosphatidylinositol biosynthesis is mediated by a complex of PIG-A, PIG-H, PIG-C and GPI1. EMBO J 17: 877-885
43. Watanabe R, Murakami Y, Mamor MD, et al (2000) Initial enzyme for glycosylphosphatidylinositol biosynthesis requires PIG-P and is regulated by DPM2. EMBO J 19: 4402-4411
44. Takahashi M, Takeda J, Hirose S, et al (1993) Deficient biosynthesis of N-acetylglucosaminyl-phosphatidylinositol, the first intermediate of glycosyl phosphatidylinositol anchor biosynthesis, in cell lines from patients with paroxysmal nocturnal hemoglobinuria. J Exp Med 177: 517-521
45. Takeda J, Miyata T, Kawagoe K, et al (1993) Deficiency of the GPI anchor by a somatic mutation of the PIG-A gene in paroxysmal nocturnal hemoglobinuria. Cell 73: 703-711
46. Miyata T, Takeda J, Iida Y, et al (1993) The cloning of PIG-A, a component in the early step of GPI-anchor biosynthesis. Science 259: 1318-1320
47. Endo M, Ware RE, Vreeke TM, et al (1996) Molecular basis of the heterogeneity of expression of glycosylphosphatidylinositol anchored proteins in paroxysmal nocturnal hemoglobinuria. Blood 87: 2546-2557
48. Noji H, Shichishima T, Saitoh Y, et al (2001) The distribution of PIG-A gene abnormalities in paroxysmal nocturnal hemoglobinuria granulocytes and cultured erythroblasts. Exp Hematol 29: 391-400
49. Rabessandratana H, Toutant J-P, Reggio H, et al (1998) Decay accelerating factor (CD55) and membrane inhibitor of reactive lysis (CD59) are released within exosomes during in vitro maturation of reticulocytes. Blood 91: 2573-2580
50. Kooyman DL, Byrne GW, McCellan S, et al (1995) In vivo transfer of GPI-linked complement restriction factors from erythrocytes to the endothelium. Science 269: 89-92
51. Sloand EM, Maciejewski JP, Dunn DE, et al (1998) Correlation of the PNH defect by GPI-anchored protein transfer. Blood 92: 4439-4445
52. Horikawa K, Nakakuma H, Kawaguchi T, et al (1997) Apoptosis resistance of blood cells from patients with paroxysmal nocturnal hemoglobinuria, aplastic anemia, and myelodysplastic syndrome. Blood 90: 2716-2722
53. Takami A, Zang W, Wang H, et al (1999) Cytotoxicity against lymphoblastoid cells mediated by a T-cell clone from an aplastic anaemia patient: role of CD59 on target cells. Br J Haematol 107: 791-796
54. Karadimitris A, Manavalan JS, Thaler HT, et al (2000) Abnormal T-cell repertoire is consistent with immune process underlying the pathogenesis of paroxysmal nocturnal hemoglobinuria. Blood 96: 2613-2620
55. Young NS, Maciejewski JP (2000) Genetic and environmental effects in paroxysmal nocturnal hemoglobinuria: this little PIG-A goes "Why? Why? Why?". J Clin Invest 106: 637-641

56. Kawagoe K, Kitamura D, Okabe M, et al (1996) Glycosylphosphatidylinositol-anchor-deficient mice: implications for clonal dominance of mutant cells in paroxysmal nocturnal hemoglobinuria. Blood 87: 3600-3606
57. Dunn DE, Yu J, Nagarajan S, et al (1996) A knock-out model of paroxysmal nocturnal hemoglobinuria: Pig-a-hematopoiesis is reconstituted following intercellular transfer of GPI-anchored proteins. Proc Natl Acad Sci USA 93: 7938-7943
58. Shichishima T, Saitoh Y, Terasawa T, et al (1997) Relationship between the phenotypes of circulating erythrocytes and cultured erythroblasts in paroxysmal nocturnal hemoglobinuria. Blood 90: 435-443
59. Hattori H, Machii T, Ueda E, et al (1997) Increased frequency of somatic mutations at glycophorin A loci in patients with aplastic anaemia, myelodysplastic syndrome and paroxysmal nocturnal haemoglobinuria. Br J Haematol 98: 384-391
60. Purow DB, Howard TA, Marcus SJ, et al (1999) Genetic instability and the etiology of somatic PIG-A mutations in paroxysmal nocturnal hemoglobinuria. Blood Cell Mol Dis 15: 81-91
61. Sugiyama H (2001) Wilms' tumor gene WT1: its oncogenic function and clinical application. Int J Hematol 73: 177-187
62. Oka Y, Elisseeva OA, Tsuboi A, et al (2000) Human cytotoxic T-lymphocytes responses specific for peptides of wild-type Wilms' tumor gene WT1 product. Immunogenetics 51: 99-107
63. Gaiger A, Reese V, Disis ML, et al (2000) Immunity to WT1 in the animal model and in patients with acute myeloid leukemia. Blood 96: 1480-1489
64. Luzzatto L, Bessler M (1996) The dual pathogenesis of paroxysmal nocturnal hemoglobinuria. Curr Opin Hematol 3: 101-110

Part 3

Clinical Pathology and Natural History of PNH

A Short History of Paroxysmal Nocturnal Hemoglobinuria: How We Came To Understand Its Natural History

Wendell F. Rosse, M.D.

Division of Hematology, Department of Medicine, Duke University, Durham, North Carolina, USA

Summary: The history of the development of the understanding of the natural history of paroxysmal nocturnal hemoglobinuria (PNH) illustrates the interaction between basic science and clinical problems. Through application of the knowledge obtained in the laboratory, a better understanding of the clinical problems has been achieved.

Key words: paroxysmal nocturnal hemoglobinuria, history, natural history, complement, stem cell disorders

Introduction

Throughout history, there have been many systems of medicine that differ from each other and from the system we use. The system of ancient Greece is well documented in the Hippocratic writings. The medical writings of India, China, Japan and other civilizations describe the concept of disease, diseases, and their treatment as viewed by those civilizations. None of these have been as successful as Western medicine in understanding and treating illness and disability.

The reason for the success of modern medicine lies in two revolutions that took place in the 19th century. The first was a revolution in the nosology of disease. Certainly in Greek medicine and through much of early Western medicine, clinical observations, often very acute ones, were made of symptoms and signs presented by the patient but they were often not assembled and named as a distinct illness. Beginning in the late 18th century, doctors began the think more in terms of diseases rather than only in terms of the various manifestations of disease. This led during the early and middle part of the 19th century to a more and more focused view of individual diseases with causes, manifestations, complications, and treatments. This more practical way of looking at illness was manifest in the texts of medicine, including the great German tracts of the 1890's and Osler's Textbook of Medicine of 1892.

The second great revolution of the 19th century was the development of basic scientific research into biological processes and their application to human disease. The first part of this revolution was part of the general burgeoning of science and was carried out, not by physicians, but by dedicated scientists working in universities and research institutes, such as the Kaiser Wilhelm Institutes in Germany and the Pasteur Institute in France. The value of this scientific knowledge in relation to human disease was recognized by the same physicians who had improved their diagnostic skill and the result was a greater ability to understand, diagnose, and treat the illnesses encountered in the clinic. This seemingly obvious apothegm – understanding basic biology is fundamental to understanding human disease – has become the basis of our reasonably successful system of medicine.

Hemoglobinuria and Hemolysis in PNH

The history of the development of knowledge about paroxysmal nocturnal hemoglobinuria illustrates well how this has come about. Although the cardinal symptom, dark urine coming on at night and clearing during the day, must have been observed many times during the passage of time, it was not recognized as a unique clinical entity until 1866, when William Gull described a leather worker with this symptom and made careful observations over a three months period [1]. He did not know what the dark pigment was but thought it might be hematin, which had just been isolated from the blood; he knew at any event that the urine did not contain red cells, only apparently the pigment from them. He noted the nocturnal and the paroxysmal pattern and related some of the paroxysms to infections. Gull himself went on to become a rich and famous physician at Guy's Hospital but never wrote another paper on a hematological subject.

Sixteen years later, in 1882, Paul Strubing, a young physician in Griefswald, Germany, a relatively obscure city on the Baltic Sea, published a brilliantly detailed account of a young army veteran with "paroxysmal hemoglobinuria" [2]. He noted the anemia, the pattern of hemoglobinuria and the symptomatology that many patients with hemoglobinuria complain of – the feeling of fatigue, the chest tightness (due to esophageal spasm), the abdominal discomfort and tenderness in the area of the spleen. He recognized the pigment as hemoglobin and distinguished the entity from other causes of hemoglobinuria known at that time – march hemoglobinuria and paroxysmal cold hemoglobinuria (PCH).

There was, at that time, a brilliant young physiologist at Griefswald, Leonhard Landois, who had been doing experiments on the effects of transfusion of blood from one species to another. Strubing and he noted similarities between the patient and the animals who had had hemolytic transfusion reactions; both had hemoglobin in the plasma as well as the urine, both had hemosiderin in the kidneys and in the urine (although the Prussian Blue test for iron in cells was not yet invented). Landois had noted thromboses in the animal model and Strubing proposed, undoubtedly correctly, that some of the symptoms and findings in his patient (e.g. the splenomegaly) could be due to clotting. Strubing correctly surmised that the hemoglobin came from the destruction of abnormal red cells in the circulation (not in the kidney), and he thought that hemolysis might result from rejection of the cells as foreign, as in Landois' xenotransfusion model. This collaboration between the observant clinician and the curious scientist set the pattern for the development of knowledge about PNH.

In 1911, a well-known physician at the Coolsingel clinic in Rotterdam, A. A. Hijmans van den Bergh, described another case of "icterus due to hemolysis" (the concept of which had been explicated by Chauffard and Trosier three years earlier) and presented some remarkable results of investigation of the "fragility of the red cells"[3]. The clinical description added little to what was already known, except to recognize the jaundice of the patient as being due to the destruction of the red cells. (Van den Bergh is best known for his chemical tests for bilirubin in the plasma.) He reasoned that the nocturnal pattern of hemoglobinuria could be due to the relative acidification that occurs due to the accumulation of CO_2 during sleep, a fact that had been observed by respiratory physiologists in the late 19th century. He mixed the cells from the patients with serum (compatible in Landsteiner's blood groups) from normal donors and from the patients themselves and incubated the mixtures under normal air and under a moderate concentration of CO_2 with "astonishing results". The patients' red cells lysed in either normal or autologous serum when incubated in CO_2!

Van den Bergh further reasoned that the lysis could be due to complement in the serum. Complement had been recognized in 1888 as a thermolabile factor in normal serum that amplified the action of what was later recognized as antibodies, both in bactericidal and phagocytic functions. Paul Ehrlich, probably the most brilliant scientist of his day, found as well that complement was needed to lyse the red cells sensitized by heterologous antibody in his experiments in xenotransfusion. Donath and Landsteiner had already applied this scientific information to the clinical problem of paroxysmal cold hemoglobinuria; they had shown that the serum of the patients contained a cold reacting antibody that, in the presence of fresh normal serum as a source of complement, lysed the red cells of both normal donors and the patients. Van den Bergh found as well that heating the serum used in his experiments to 56. C abrogated lysis of the patient's red cells; however, he could not restore the activity by adding back fresh normal serum, probably because, as we now know, very high concentrations of complement are needed and the reaction mixture was too dilute. Little more was known of complement except its thermolability and he could not investigate further its possible role, missing an opportunity to open the field. He is, however, the first physician investigating PNH who applied the findings and techniques of basic science to try to understand clinical facts, a template that would be followed by many after him.

During the next 27 years, more clinical descriptions of PNH appeared. The variability and chronicity of the syndrome was well established. A number of papers described thrombosis, particularly of the hepatic veins, a finding more thoroughly defined later by Hartmann and his associates [4]. The syndrome got its name but little was done to try to explain the clinical findings until 1938 when three papers appeared, all showing the same phenomenon that van den Bergh had adumbrated. The first was that of his pupil, Jordan, who in is thesis paper three years earlier had shown that acidification of the serum in which PNH red cells were incubated caused them to be lysed [5]. He felt that the lysis was due to complement, which had become better defined by this time but could not define an antibody, thought to be necessary for the activation of complement, and assumed that the cells had been sensitized in vivo. Thomas Hale Ham, an American hematologist and colleague of William Castle, presented similar data later that same year and showed further that the red cells in PNH were sensitive to lysis by complement no matter how it was activated [6]. He too was perplexed by the absence of evidence of an antibody to activate the complement in acidified normal serum, as was Dacie, a young English pathologist who published similar data that same year [7]. Although all three were puzzled by the result, they clearly established a diagnostic test since the red cells of patients with no other disease were lysed by acidified normal serum (much later, it was shown that the red cells of patients with a congenital dyserythropoietic disorder were also lysed in this test but the mechanisms proved to be much different). This permitted the diagnosis of the disease even though the cardinal symptom was absent.

The characteristics of complement were slowly being revealed. As early as 1911, it was shown to consist of at least two components since one (called the first component of complement or C'1) precipitated out in the euglobulin fraction of serum and the other in the pseudoglobulin fraction (these fractions were generated by reducing the ionic strength of the serum); this component was shown to need calcium for its activity. A second component (C'2) required magnesium for its activity and another component (C'4) was inactivated by hydrazine and other similar agent. The fraction needed for lysis of cells was designated C'3. Antibody was found to bind C'1 which then fixed C4 to the cell. C'1 also activated C2 and the combination of C4 and C'2 activated C'3 [8].

This simple view was challenged by Pillemer and his associates found that an extract of the yeast cell wall, zymosan, could activate complement without the intervention of C'1, C'2, or C'4 and that this activation involved other proteins, including one which he called "properdin" [9]. This alternative activation of complement offered a solution to the problem of how PNH cells could be lysed without antibody; Hinz, a clinician colleague of Pillemer, demonstrated this using the techniques and reagents of Pillemer 10]. However, the concept of an alternative pathway of complement activation was not accepted by the immunological establishment; Pillemer became despondent that he could not get his work accepted and published and he committed suicide (a possibly apocryphal story says he did so by drinking veronal buffer which was frequently used in complement research). One finding that the critics of Pillemer could not explain by any other means was the lysis of PNH red cells by complement when all evidence of antibody was lacking.

Pillemer had had to use the techniques available to him to distinguish different proteins of complement but in the late 1950's and early 1960's, separation and purification techniques, particularly chromatography, became available. It was quickly shown that C'3 was actually six proteins that acted sequentially to produce the membrane lesion that lysed the cell. Further, Pillemer's work was confirmed and the proteins that acted in the alternative pathway were isolated and shown to activate C3 (in the new nomenclature). This activation by acidification of the normal serum is sufficient to lyse PNH red cells; the problem that had perplexed Ham, Jordan and Dacie was solved.

These studies did not show why PNH red cells were so much more sensitive than normal cells to the hemolytic action of complement. Careful analysis had shown that most patients had a population of cells that required about 1/25 the concentration of serum as normal cells (a few patients had cells that required only 1/5 or so as much serum as normal cells for equal lysis [11]. Because PNH cells were readily lysed by the activation of the alternative pathway, it appeared likely that the steps that enhanced lysis would occur at the steps involving C3 or beyond. Charles Parker made the important observation that much more C3 was laid down on the surface of a PNH cell than on a normal cell when a same number of enzymatic units (C3 convertase) were present [12]; it was known that this enzyme unit consists of a molecule of C4 (or activated C3, C3b, in the alternative pathway) attached to the membrane and an activated enzyme, C2b (classical pathway) or Bb (alternative pathway) liganded to it by magnesium. This unit is intrinsically unstable and when the two elements are separated, they can no longer activate their substrates, C3 and C5. Edward Hoffman, an immunologist in Florida, had isolated a substance which increased the rate of dissociation of the two components of the convertase and which he called DAF or "decay accelerating factor"[13]. Although initial studies failed to demonstrate a difference between normal and PNH cells with respect to this factor, more careful and detailed studies in 1983 by a Boston clinician-scientist, Anne Nicholson-Weller, first demonstrated that PNH red cells lacked this protein [14]. Thus, the results of Parker could be explained by the greater stability of the C3 convertase complexes on the cell surface that were then able to activate more C3.

It was initially thought that the lack of DAF was sufficient to account for the increased sensitivity of PNH red cells to the lytic action of complement. However, inhibition of DAF on normal cells by specific antibodies did not render those cells PNH-like. Further, less cell-bound C3 was needed to lyse PNH cells than normal cells, suggesting that some later step was also abnormal. This abnormality was identified in 1989 by Parker when he isolated a protein from normal red cells that inhibited the lysis of PNH red cells in acidified serum; he called the protein "membrane inhibitor of reactive lysis" or "MIRL"[15] but if turned out to be the same as a previously identified lymphocyte entity called CD59 and the latter designation is usually used. It inhibits the formation of the ultimate membrane attack complex, the polymeric C9 complex that is capable of disrupting the lipid bilayer and of causing lysis. CD59 is lacking on PNH red cells and its inhibition on normal cells results in the PNH phenotype with respect to complement lysis. Again, basic understanding of the workings of complement had solved the problem of the cause of hemolysis in PNH.

In the patient with PNH, hemolysis occurs whenever complement is activated. The nocturnal pattern is probably due to the circadian absorption of endotoxin from the gut as endotoxin is a potent activator of the alternative pathway. Almost any infection will activate complement, as will some immunizations. The amount of hemolysis depends not only on the degree to which complement is activated but also on the proportion of abnormal cells and their abnormality. Considerable hemolysis is needed before hemoglobinuria becomes evident since much of the hemoglobin that is filtered through the glomerulus of the kidney is reabsorbed in the proximal tubule, leading acutely to a Fanconi's like abnormality of proximal tubular function and chronically to proximal tubular acidosis or even renal failure in about 5% of patients.

The free hemoglobin also apparently causes symptoms. Hemoglobin readily binds nitric oxide (NO), thus removing it from tissues. This is inhibited by the isolation of hemoglobin in cells but occurs readily when free hemoglobin bathes the cells. NO causes smooth muscle relaxation and its removal results in excessive contraction of smooth muscles. This is especially notable in the esophagus; many patients with PNH and hemoglobinuria have esophageal spasm during the hemoglobinuria (first noted by Strubing as "chest pain"), which can be relieved by nitroglycerine. Almost all men with PNH are sexually impotent during periods of hemoglobinuria, presumably due to contraction of arteries supplying the corpora cavernosa of the penis; this symptom is sometimes relieved by assuring a supply of NO. Again, basic knowledge of NO biology helped in the understanding of clinical problems of PNH.

Thrombosis in PNH

Almost from the first descriptions, the increased occurrence of thromboses has been noted among patients with PNH. Peculiarly, these thromboses appeared to occur primarily in veins of the abdomen, particularly the hepatic veins. Other sites included the cerebral veins and the veins of the extremities. Even the dermal veins were sometimes affected.

Curiously, hypercoaguability appears to be much more common in patients of European origin than in patients of East Asian origin. In several studies from Europe and the United States, about 40% of patients have thrombosis at some time or another. In China, Japan, Thailand and Mexico, only 5-10% of patients suffer this complication (it should be remembered that about 75% of the gene pool in Mexico is native American and is thus derived from Asian sources). The reason for this difference is not known.

William Crosby, writing in the early 1950's emphasized the lethality of these complications and, from the fact that thrombosis often occurred during a paroxysm of hemolysis; he proposed that the activation of the coagulation system activated complement [16]. At that time, detailed knowledge of either the coagulation system or of the complement system was not available and he was never able to establish satisfactorily that either system activated the other.

When it was recognized that the components of the coagulation system in plasma used membranes in the initiation of their activation, it was proposed that the release of the membranes from lysed red cells might activate coagulation; however, in vitro evidence was equivocal and other settings in which intravascular hemolysis occurred do not result in the same degree of increased thrombosis. The genesis of the tendency to thrombosis was not understood.

In 1969, Enright and Aster demonstrated that the platelets (and neutrophils) shared with the red cells sensitivity to the lytic action of complement, at least in vitro cytolytic assays [17]. Many years later, both DAF and CD59 were found to be deficient on a population of platelets in all patients and several studies suggested that these platelets were more prone to aggregation than normal platelets. But how these facts related to the thrombotic tendency was not clear.

At least part of the solution came from the observation that, in the absence of splenomegaly, platelets in PNH, unlike the abnormal red cells, had a normal survival in the circulation, despite the deficiency in the complement defense proteins [18]. Part of the explanation for this appeared the be the ability of the activated platelet to secrete factor H of the complement system, which acts much like DAF in down-regulating the C3 convertase, from the internal granules. However, this did not appear to be sufficient to account for the ability of platelets to resist complement activation.

Sims and his associates had shown that normal platelets are able nascent polymeric C9 complexes generated by the activation of complement into a "patch" and to excise them by forming an exocytic vesicle, in this way avoiding the accumulated membrane damage that these complexes mediate [19]. However, these vesicles are not able to maintain phospholipid "sidedness" and the acidic phospholipids, particularly phospatidylserine, appear on the surface in contact with plasma. There, they serve as sites for the deposition of the coagulation factors that constitute a prothrombinase complex; thrombin is generated which fosters clotting and which aggregates platelets. The result is a hypercoaguable state.

Weidmer and her associates found that these reactions are markedly exaggerated on PNH platelets. Because the formation of polymeric C9 complexes is not inhibited by CD59, many exocytic vesicles are formed with even minimal amounts of complement activation [20]. Thrombin is generated and, since the PNH platelets are unusually susceptible to aggregation by thrombin, more aggregation results. These aggregates presumably form the nidus for thrombus formation.

The regulation of the clotting system is so complex that there may be other reasons not yet elucidated for the hypercoaguable state in PNH. For example, since PNH monocytes lack the receptor for plasmin activator [21], fibrinolysis may also be diminished, increasing the chances of significant clot formation. Whatever the reasons, the hypercoaguable state in PNH is a major cause of morbidity and mortality and the history of thrombosis, particularly intraabdominal thrombosis, is a significant negative prognostic finding.

The PNH Stem Cell

The concept of an hematopoietic stem cell, that is, a cell which could give rise to any of the hematopoietic elements upon differentiation, is a very old one and probably dates back to the speculations of Bloom in the 1920's. However, little real evidence that such a cell exists did not come until the experiments of Jacobson in the late 1940's in which he injected marrow into lethally irradiated mice and analyzed the colonies that grew in the spleen [22]. As Till and McCollough later showed, most of the colonies were mixed and contained precursors of red cells, white cells and platelets [23]. Since each colony was apparently derived from a single cell, the conclusion was drawn that a sufficiently primitive cell could give rise to all three types of hematopoietic cells.

Attempts to isolate the hematopoietic stem cell have been consistently unsuccessful until very recently so that the evidence for its existence remained indirect. Clinically, it was clear that all three cell lines were disturbed in the myeloproliferative and myelodysplastic disorders, suggesting that the defect originated in the stem cell. Where somatic mutations could be detected, as in the case of the Philadelphia chromosome of chronic granulocytic leukemia, it was found in all the cells apparently derived from the marrow (red cells, granulocytes and monocytes, platelets and lymphocytes) shared the defect.

The finding by Enright and Aster that, in PNH, both neutrophils and platelets were, like red cells, susceptible to complement lysis implied that the defect was present in the stem cells and was passed to their progeny. Further, Oni and Luzzatto had shown earlier that, in a patient heterozygous for an abnormal glucose-6-dehydrogenase, all of the red cells with the PNH defect bore the same type of enzyme [24]. Other polymorphic markers on the X chromosome were also found to be present in a single phenotype on the abnormal cells, again indicating their clonal origin. This data was interpreted to mean that a single defective stem cell gives rise to the abnormal population of hematopoietic cells characteristic of the syndrome. PNH thus joined the other "clonal" disorders of the hematopoietic system.

In almost all patients, cells derived from normal clones of stem cells are present in marrow and blood. The proportion of abnormal cells ranges from 2-3% to, most commonly, over 80%. The severity of many of the manifestations of PNH depends upon the proportion of the blood cells that are abnormal.

The nature of the defect was not at all clear. By the late 1980's, at least four different proteins were known to be missing on PNH cells. William Beck has first shown that some of the neutrophils completely lacked the membrane enzyme, alkaline phosphatase, whereas the remaining cells had apparently normal amounts of the enzyme [25]. Hartmann and Auditore had demonstrated that the red cells had diminished amounts of acetylcholinesterase; the enzyme was later shown to be completely lacking on the most abnormal cells and diminished on those less abnormal [26]. These enzymes and the missing complement defense proteins, DAF (CD55) and CD59, were not all on the same chromosome so a single large somatic deletion of genetic material could not account for their absence on the PNH cells.

Again, the answer came from unrelated basic research. Low had discovered in 1977 that alkaline phosphatase could be removed intact and active from its membrane site by a bacterial enzyme specific for cleaving phosphatidyl inositol, phosphatidylinositol specific phospholipases C (PIPLC); this implied that the protein had been attached to the membrane by an interaction with phosphatidyl inositol [27]. The nature of the anchor was uncovered by analysis of Thy-1, a lymphocyte protein similarly removed by PIPLC. In addition to phosphatidyl inositol, it contained a molecule of N-glucosamine, three molecules of mannose, and one or more molecules of phosphoethanolamine linearly attached. The protein was attached by its carboxyl end to the phosphoethanolamine and the anchor and protein were attached to the membrane by the insertion of the phosphatidyl inositol into the lipid bilayer [28].

The glycosylphosphatidyl inositol (GPI) anchor was used by very primitive organisms and its biosynthetic sequence was worked out in trypanosomes. It is synthesized linearly on the outer surface of the endoplasmic reticulum, starting with PI to which is added N-acetylglucosamine (which is subsequently deacetylated), followed by three mannoses and finally the PEA. At some point late in its synthesis, it is internalized into the endoplasmic reticulum. There it is attached to nascent proteins by a transamidase and the anchored proteins were then exported through the Golgi apparatus to the external membrane.

A series of lymphocyte cell lines that did not express Thy-1 (and other GPI-anchored proteins) were developed by Tartikoff and his associates. These lines were originally classed by their ability or inability to complement one another when somatically fused and were designated as A-H. When the biosynthetic intermediates in each line were analyzed, it was found that three of them (A, C and H) did not express any intermediates, suggesting that at least three components were needed for the initial step of biosynthesis, the addition of N-acetylglucosamine to PI. Analysis of the intermediates of the biosynthesis of the anchor in PNH cells found that that the defective step in synthesis was also at this initial step and somatic fusion studies demonstrated that the defect was the same as that in the line designated A. With that information, Kinoshita and his associates, in a monumentally important series of studies, used that cell line to isolate and identify the gene defective in PNH, the pig-A gene. Once again, basic science had shed great light upon PNH [29].

Hematopoiesis in PNH

Even in very early descriptions, a relationship between PNH and marrow hypoplasia had been noted but Lewis and Dacie, in 1963, clearly iterated this relationship [30]. They pointed out that many patients with PNH had had aplastic anemia preceding it; in later studies, when aplastic anemia became more amenable to treatment, as many as 35% of patients recovering from aplastic anemia had evidence of PNH cells at some time in their blood or marrow. In some cases, the number of PNH cells remains small but in others, they become predominant and the patient has the clinical complications of PNH.

Lewis and Dacie also noted that some patients with PNH became aplastic during the course of their disease. This is clearly an adverse prognostic sign. There were indications as well that even in patients with a cellular bone marrow, hematopoiesis was suboptimal. The number of reticulocytes was less than expected. Granulocytopenia and thrombocytopenia were both common in patients with PNH; since, in the absence of splenomegaly, the survival of these cells in the circulation is normal, their production must be decreased to account for these deficiencies. When in vitro culture methods became available, they demonstrated that the proliferation of cells from PNH marrow (both normal and GPI-) was markedly less than from normal marrow. Finally, a few patients (1-3%) developed acute (usually non-lymphocytic) leukemia as a complication of PNH [31]; in each case, the leukemic cells were derived from the GPI- clone. All of these facts suggest that hematopoiesis is abnormal in PNH.

This gives rise to a paradox: :If hematopoiesis is decreased in PNH, how does the abnormal clone become dominant, often consisting of more than 80% of the hematopoietic precursors? The answer is far from clear but an hypothesis (sometimes called the Young-Luzzatto) hypothesis has been enunciated:

1. GPI (-) precursors commonly occur in otherwise normal marrow in very small numbers. Several recent studies have shown that this is probably true.
2. an separate and unrelated (perhaps immunological) event suppresses the marrow in a form of hypoplastic or aplastic anemia
3. the normal cells are suppressed more than the PNH cells, which are then enriched. This could be due to the lack of GPI-anchored receptors for suppressive influences on the PNH cells.

The data in favor of this hypothesis is being accumulated but the fact remains, we do not have a sufficient understanding of the factors controlling growth and differentiation of the marrow. If history is true to itself, this basic understanding of biology will come about and will be applied to the understanding of PNH.

Summary

The history of the development of knowledge of PNH provides a superb example of why modern medicine has made the remarkable progress it has over the past 125 years. The increasingly accurate diagnosis and clinical observations have posed questions to curious physicians. The answers have come from the application of the techniques and findings of basic biology to these clinical problems. While this interaction seems natural to us who have been raised understanding it, it is not the only way that things might have been done. We need to be thankful to those who, in the late 19th and early 20th century, directed modern medicine toward this rewarding and continuing interaction.

References

1. Gull WW (1866) A case of intermittent haematinuria, with remarks. Guy's Hospital Reports, 12:381-392.
2. Strubing P (1882) Paroxysmale Hämoglobinurie. Dtsch. Med. Wochenschr., 8:1-8.
3. Hijmans van den Bergh AA (1911) Ictère hemolytique avec crises hemoglobinuriques. Fragilité globulaire. Rev.Médecine, 31:63-69.
4. Peytremann R, Rhodes RS, Hartmann RC (1972) Thrombosis in paroxysmal nocturnal hemoglobinuria (PNH) with particular reference to progressive, diffuse hepatic venous thrombosis. Series Haematologica., 5: 115-136.
5. Jordan FLJ (1938) Études sur l'hémoglobinurie. Acta Med.Scandinav., 95:319-340.
6. Ham TH (1938) Chronic hemolytic anemia with paroxysmal nocturnal hemoglobinuria. A study of the mechanism of hemolysis in relation to acid-base equilibrium. New Engl.J.Med., 217:915-917.
7. Dacie JV, Israëls MCG, Wilkinson JF (1938) Paroxysmal nocturnal hemoglobinuria of the Marchiafava type. Lancet 1:479-482.
8. Ross GD (1986) "Introduction and history of complement research" in Immunobiology of the Complement System. Ed. G D Ross Academic Press, Orlando. P.1.
9. Pillemer L, Blum L, Lepow IH, et al. (1954) The properdin system and immunity: I. Demonstration and isolation of a new serum protein, properdin, and its role in immune phenomena. Science, 120:279
10. HinzCJ jr, Jordan WS, Pillemer LT (1956) The properdin system and immunity. IV. The hemolysis of erythrocytes from patients with paroxysmal nocturnal hemoglobinuria. J.Clin.Invest., 35:453-453.
11. Rosse WF, Dacie JV (1966) Immune lysis of normal human and paroxysmal nocturnal hemoglobinuria red blood cells. I. The sensitivity of PNH red cells to lysis by complement and specific antibody. J.Clin.Invest., 45:736-748.
12. Parker CJ, Baker PJ, Rosse WF. (1982) Increased enzymatic activity of the alternative pathway convertase when bound to the erythrocytes of paroxysmal nocturnal hemoglobinuria. J. Clin. Invest. 69:337-346

13. Hoffman EM (1969) Inhibition of complement by a substance isolated from human erythrocytes. II. Studies on the site and mechanism of action. Immunochemistry, 6:405-419.
14. Nicholson-Weller A, March JP, Rosenfeld SI et al. (1983) Affected erythrocytes of patients with paroxysmal nocturnal hemoglobinuria are deficient in the complement regulatory protein, decay accelerating factor. Proc.Natl.Acad.Sci.USA, 80:5430-5434.
15. Holguin MH, Frederick LR, Bernshaw NJ, et al. (1989) Isolation and characterization of a membrane protein from normal human erythrocytes that inhibits reactive lysis of the erythrocytes of paroxysmal nocturnal hemoglobinuria. J.Clin.Invest., 84:7-17.
16. Crosby WH, Dameshek W (1950) Paroxysmal nocturnal hemoglobinuria. The mechanisms of hemolysis and its relation to the coagulation mechanism. Blood, 5:822-842.
17. Aster RH, Enright SE (1969) A platelet and granulocyte membrane defect in paroxysmal nocturnal hemoglobinuria: Usefulness for the detection of platelet antibodies. J.Clin.Invest., 48:1199-1210.
18. Devine DV, Siegel RS, Rosse WF (1987) Interactions of the platelets in paroxysmal nocturnal hemoglobinuria with complement. Relationship to defects in the regulation of complement and to platelet survival in vivo. J.Clin.Invest., 79:131-137.
19. Sims PJ, Faioni EM, Wiedmer T et al. (1988) Complement proteins C5b-9 cause release of membrane vesicles from the platelet surface that are enriched in the membrane receptor for coagulation factor Va and express prothrombinase activity. J.Biol.Chem., 263:18205-18212.
20. Wiedmer T, Hall SE, Ortel TL, et al. (1993) Complement-induced vesiculation and exposure of membrane prothrombinase sites in platelets of paroxysmal nocturnal hemoglobinuria. Blood, 82:1192-1196.
21. Ploug M, Plesner T, Ronne E, et al. (1992) The receptor for urokinase-type plasminogen activator is deficient on peripheral blood leukocytes in patients with paroxysmal nocturnal hemoglobinuria. Blood, 79:1447-1455.
22. Jacobson LO, Simmons EL, Marks EK, et al. (1951) Further studies on recovery from radiation injury. J Lab Clin Med 37:683.
23. Till JF, McCollouch EA (1961) A direct measurement of the radiation sensitivity of normal mouse bone marrow cells. Radiation Res. 14:213
24. Oni SB, Osukoya BO, LuzzattoL (1970) Paroxysmal nocturnal hemoglobinuria: Evidence for monoclonal origin of abnormal red cells. Blood 36:145.
25. Beck WS, Valentine WN (1951) Biochemical studies on leucocytes. II. Phosphatase activity in chronic lymphatic leukemia, acute leukemia, and miscellaneous hematologic conditions. J.Lab.Clin.Med., 38:245-253.
26. Auditore JV, Hartmann RC, Flexner JM et al. (1960) The erythrocyte acetylcholinesterase enzyme in paroxysmal nocturnal hemoglobinuria. Arch.Pathol., 69:534-540.
27. Low MG, Finean JB (1977) Release of alkaline phosphatase from membranes by a phosphatidylinositol-specific phospholipase C. Biochem.J. 167:281-284.
28. Tse AGD, Barklay AN, Watts A, et al. (1985) A glycophospholipid tail at the carboxyl terminus of the Thy-1 glycoprotein of neurons and thymocytes Science 230:1003
29. Takahashi M, Takeda J, Hirose S, et al. (1993) Deficient biosynthesis of N-acetylglucosaminyl- phosphatidylinositol, the first intermediate of glycosyl

phosphatidylinositol anchor biosynthesis, in cell lines established from patients with paroxysmal nocturnal hemoglobinuria. J.Exp.Med. 177:517-521.

30. Dacie JV, Lewis SM (1961) Paroxysmal nocturnal haemoglobinuria: variation in clinical severity and association with bone marrow hypoplasia. Br.J.Haematol., 7:442-457.
31. Harris JW, Koscick R, Lazarus HM (1999) Leukemia arising out of paroxysmal nocturnal hemoglobinuria Leukemia and Lymphoma 32:401-426

Clinical Pathology and natural history of PNH; The French Society of Hematology experience.

Gérard Socié[1], M.D., Ph.D., and Jean-Yves Mary[2], Ph.D

[1]Service d'Hématologie-Greffe de Moelle, Hôpital Saint Louis, 1 Avenue Claude Vellefaux 75475, Paris CEDEX 10, France
[2]Centre de Bioinformatique, INSERM U444, Université Paris VII, 2 Place Jussieu, 75251 Paris CEDEX 05, France

Summary: Paroxysmal Nocturnal Haemoglobinuria (PNH) is a rare acquired hematopoietic stem cell disorder caused by *PIG-A* gene alterations. In contrast to the growing knowledge regarding the pathophysiology of the disease, data on factors influencing survival are scarce, mainly as a result of the rarity of the disease. Data on 220 patients, diagnosed with PNH over a 46-year period in reporting French centers, were collected. Kaplan-Meier survival estimates (mean +/- standard error) were 65±4% and 48±6% at 10 and 15 years after diagnosis, respectively. In this cohort of 220 patients with PNH, aplastic anemia (AA) antedated PNH in 65 patients with a median interval of 3.1 years. These 65 patients with an AA-PNH syndrome were shown to have a relative risk (RR) of death 3 times lower than the remaining 155 patients with *de novo* PNH. Finding a better prognosis among patients with AA before diagnosis of PNH might be surprising, we thus reassessed, the differences between the AA-PNH syndrome and *de novo* PNH. Age, gender and symptom distributions were similar in both groups, at diagnosis. However, when comparing these patients at presentation, anemia was less frequently, but thrombopenia and neutropenia were more frequently observed. Patients with the AA-PNH syndrome were more frequently treated initially with androgens or immunosuppressive therapy than those with *de novo* PNH. During follow-up, the prevalence or cumulative incidence of main complications was similar in both groups. Multivariate analysis showed that 5 factors influenced survival in both groups (thrombosis, MDS or acute leukemia, age over 54 years, thrombocytopenia at diagnosis and initial treatment other than androgens), while 1 factor (pancytopenia during the evolution) increased the risk of death in patients with *de novo* PNH. Treatment other than androgens led to an increased risk of death (~ 5 fold) in patients with *de novo* PNH while it lowered the risk of death (~ 4 fold) in patients with the AA-PNH syndrome, compared to patients treated by androgens in either group.

Key words: Aplastic anemia, paroxysmal nocturnal hemoglobinuria, long-term

Introduction

In paroxysmal nocturnal hemoglobinuria (PNH), the marrow function is frequently found to be impaired, as shown by pancytopenia and a relatively low reticulocyte count in light of the degree of anemia (review in: [1]). In a number of patients, moderate to severe marrow failure may occur during the course of the disease. Another group of patients has long been recognized, in whom the first diagnosis is aplastic anemia (AA), and the signs of PNH have become manifest later. Thus, from a clinical point of view, PNH may be considered a disease presenting two forms: *de novo* PNH and the AA-PNH syndrome [2]. Although the AA-PNH syndrome has been recognized since the mid 40's (early references are reviewed by Lewis & Dacie in [3]), no large scale studies comparing patients with *de novo* PNH and with the AA-PNH syndrome have been conducted nowadays, most likely because the disease is so rare.

PNH long-term follow-up and prognostic factors

Data on 220 patients, diagnosed with PNH over a 46-year period (1950-1995) in reporting French centers, were collected [2]. Diagnosis of the disease was made by, at the least, an unequivocally positive Ham's test. Kaplan-Meier survival estimates (mean +/- standard error) were 65±4% and 48±6% at 10 and 15 years after diagnosis, respectively. Eight-year cumulative incidence rates of main complications, i.e.; pancytopenia, thrombosis and myelodysplastic syndrome were 15±3%, 28±4% and 5±2%, respectively (**Figure 1**). Demographic, presenting features, initial treatment, complications and causes of death data were first analyzed and were found to be similar to those generally reported in the literature. This large number of cases permitted a detailed analysis of prognostic factors for the first time, in this rare disease. Using multivariate analysis, seven factors were shown to be significantly associated with survival in patients with PNH. Poor survival was associated with the occurrence of thrombosis as complication [Relative Risk (RR)=10.2 CI=6-17, p<.0001], evolution to pancytopenia [RR=5.5 (2.8-11), p<.0001]], myelodysplastic syndrome or acute leukemia [RR=19.1 (7.3-50), p<.001], age over 55 years at diagnosis [RR=4 (2.4-6.9), p<.0001], need for additional treatment [RR=2.1 (1.3-3.6), p<.003], and thrombocytopenia at diagnosis [RR=2.2 (1.3-3.8), p<.003]. Better survival was shown for patients in whom aplastic anemia antedated PNH [RR=0.32 (0.14-0.72), p<.02]. Furthermore, we analyzed presenting features, at diagnosis, that may influence severe PNH events that occurred during evolution (i.e., thrombosis, myelodysplastic syndrome or acute leukemia, and evolution to pancytopenia), by multivariate analysis. Factors associated with a high risk for thrombosis were: thrombosis at diagnosis

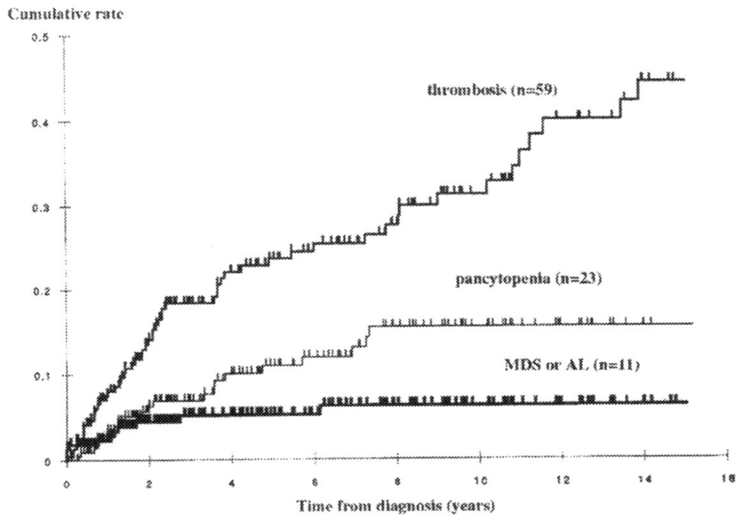

Fig. 1. Complication rates in PNH

[RR=5.1 (2.5-10.6), p=.0002], age over 54 years [RR=2.6 (1.5-4.6), p=.0014] and, infection at diagnosis [RR=2.6 (1.3-5.2), p=.0099]. The risk factors for the evolution to pancytopenia were: absence of anemia at diagnosis [RR=4.03 (1.3-12.2), p=.03] and neutropenia at diagnosis [RR=2.45 (1.1-5.7), p=.03]. The risk factors for the evolution to myelodysplastic syndrome or acute leukemia were: abdominal pain crisis at presentation [RR=10.5 (2.5-44.0), p=.004] and year of diagnosis beyond 1983 [RR=8.45 (1.8-40.7), p=.004].

The Aplastic anemia / PNH syndrome

We collected data on 220 patients diagnosed with PNH between 1950 and 1995. In a first study we described the long-term follow-up and prognostic factors in this patient-cohort. We found that patients with the AA-PNH syndrome have a better prognosis (3 times less risk of death) than patients with *de novo* PNH [2]. Finding a better prognosis among patients with AA before diagnosis of PNH might be surprising. Consequently, on this large group of patients, our aim was to reassess the differences between the AA-PNH syndrome and *de novo* PNH.

Biological and clinical symptoms at diagnosis and initial treatment are summarized in **Table 1**. Age, gender and symptom [thrombosis, infection and abdominal pain crisis] distributions were similar in both groups, at diagnosis.

Table 1. Patient characteristics.

Variable	Category	Total [# 220]	AA-PNH [# 65]	*de novo* PNH [# 155]	p* value
Gender; n, %	Female	120, 55%	37, 57%	83, 54%	.65
Age(year): mean ± SD	NA	37 ± 18	37 ± 18	38 ± 18	.72
Thrombosis; n, %	Yes	14, 6%	2, 3%	12, 8%	.20
Abdominal pain crisis; n, %	Yes	28, 13%	4, 6%	24, 15%	.06
Infection; n, %	yes	18, 8%	4, 6%	14, 9%	.48
Anemia; n, %	yes	196, 89%	52, 80%	144, 93%	.005
Neutropenia; n, %	yes	90, 41%	34, 52%	56, 36%	.03
Thrombocytopenia; n, %	yes	112, 51%	41, 63%	71, 46%	.02
Pancytopenia; n, %	yes	70, 32%	26, 60%	44, 72%	.09
Treatment; n, %					
Androgens	yes	62, 28%	34, 52%	28, 18%	<.0001
Danazole	yes	11, 5%	4, 6%	7, 5%	.61
Corticosteroids	yes	45, 20%	16, 25%	29, 19%	.32
CSA and/or ATG	yes	21, 10%	14, 22%	7, 5%	.0001
BMT	yes	4, 2%	0, 0%	4, 3%	.24
Transfusions	yes	164, 75%	44, 68%	120, 77%	.13
Others	yes	13, 6%	1, 2%	12, 8%	.08

CSA; cyclosporine A. ATG; antithymocyte globulin.
BMT; bone marrow transplantation

However, when comparing patients with the AA-PNH syndrome to patients with *de novo* PNH at presentation, anemia was less frequently observed (80 vs. 93%), while thrombopenia and neutropenia were more frequently detected (63% vs. 46% and 52% vs. 36%, respectively). For patients with anemia, the mean (SD) hemoglobin level was 8.1 (1.8) g/L and 8.0 (1.8) g/L in the AA-PNH syndrome and in the *de novo* PNH groups, respectively [p=.83]. The corresponding figures for neutropenia and thrombocytopenia were 0.8 (0.4) $x10^9/L$ vs. 0.9 (0.4) $x10^9/L$ [p=.052] and 39.9 (4.0) $x10^9/L$ vs. 48.8 (30.5) $x10^9/L$ [p=.17], respectively. Patients with the AA-PNH syndrome were treated initially with androgens (52% vs. 18%) or with immunosuppressive therapy (22% vs. 5%) more frequently than patients with *de novo* PNH.

During follow-up, the prevalence of infections or abdominal pain crises, cumulative incidences of myelodysplastic syndrome or acute leukemia, and the cumulative proportion of patients requiring additional treatment were similar in both groups. The cumulative incidence of thrombosis (or life-threatening thrombosis) was slightly lower in patients with the AA-PNH syndrome as compared to patients with *de novo* PNH, **Table 2**.

Specific causes of death in both groups are summarized in **Table 3**. It reveals substantial differences in the causes of death in the two groups of patients. A

Table 2. Complications occurring overall & in both groups of patients.

Complications	**Total # 220**	**AA-PNH # 65**	***de novo* PNH; # 155**	**P value**
MDS or AL				
n (%)	11 (5%)	4 (6%)	7, (5%)	
Time (year) since diagnosis :mean ± SD	1.4 ± 1.8	0.7 ± 0.9	1.7 ± 2.2	
10-yr incidence (%) : mean ± SD	6 ± 2	7 ± 3	6 ± 2	.61
Recurrent abdominal pain crisis, n (%)	41 (19%)	12 (18%)	29 (19%)	.95
Recurrent Infections, n (%)	41 (19%)	10 (15%)	31 (20%)	.40
Pancytopenia	NA	NA		NA
n (%)			23 (15%)	
Time (year) since diagnosis : mean ± SD			3.4 ± 3.8	
10-yr incidence (%) : mean ± SD			22 ± 4	
Any Thrombosis				
n (%)	58 (26%)	11 (17%)	47 (30%)	
Delay (year) since diagnosis : mean ± SD	4.4 ± 5.1	3.4 ± 3.6	4.7 ± .5.4	
10-yr Cumulative incidence (%) : mean ± SD	32 ± 4	25 ± 8	34 ± 5	.16
Other; n (%)	40 (18%)	14 (22%)	26 (17%)	.40
Hemochromatosis; n	11	3	8	
Hepatitis B or C; n	11	5	6	
Malignant disorders; n	9	4	5	
Other; n	9	2	7	

Table 3. Primary causes of death.

Cause of death (patients who underwent BMT, excluded)	**Total # 62: n (%)**	**AA-PNH # 6: n (%)**	***de novo* PNH # 56: n (%)**
Budd-Chiari syndrome	15 (24%)	0 (0%)	15 (27%)
CNS vascular complications	18 (29%)	2 (33%)	16 (29%)
thrombosis	*8*	*1*	*7*
hemorrhage	*6*	*1*	*5*
unknown	*4*	*0*	*4*
Infectious diseases	12 (19%)	2 (33%)	10 (18%)
Malignancies	4 (6%)	0 (0%)	4 (7%)
Including 3 deaths from MDS/AL			
Other causes	13 (21%)	2 (33%)	11 (20%)
Unknown cause	*6*	*1*	*5*
*Disease or treatment related**	*6*	*1*	*5*
Disease or treatment unrelated †	*1*		*1*

* 3 cases of hemorrhage as consequence of pancytopenia in patients with *de novo* PNH, 2 of hemochromatosis (1 in each group), 1 of acute renal insufficiency (hemolysis in a patient with *de novo* PNH)

† 1 case of myocardial infarction.

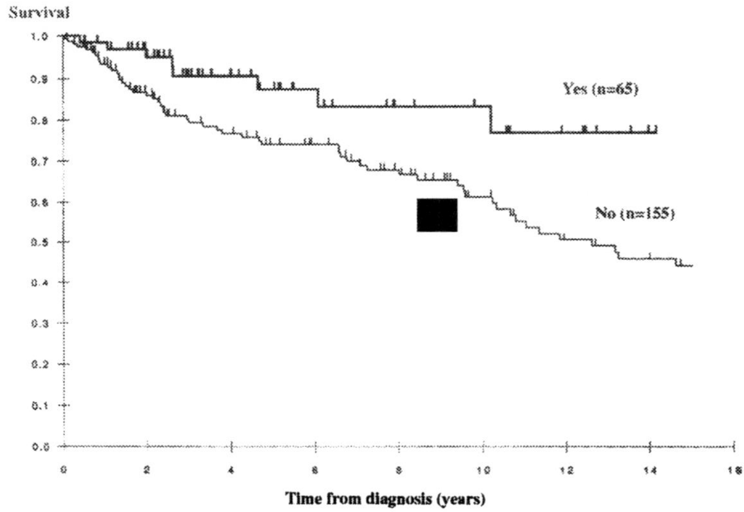

Fig. 2. Survival according to the occurrence of a previous aplastic anemia.

Budd-Chiari syndrome was the primary cause of death in 24% of the *de novo* PNH patients, while none of the 6 patients with the AA-PNH syndrome died of this complication (p=.17). Similarly, CNS thrombosis accounted for 8 death overall [1 in the AA-PNH syndrome group and 7 in the *de novo* PNH group (p=.49)].

Demographic data, initial signs and symptoms, initial treatment type, complications occurring during evolution, necessity and type of subsequent treatment (if any) were analyzed, in relation to survival, one by one in both groups separately. Among the 25 factors tested in univariate analysis, 7 factors appeared to be significantly related to poor survival, at least in one group. These relations are the followings in the AA-PNH syndrome group and in the *de novo* PNH group, respectively: thrombosis at presentation [RR=39.1, p<.0001 and RR=2.0, p=.09], initial treatment without androgens [RR=0.32, p=.27 and RR=3.9, p=.014], thrombosis during evolution [RR=76.6, p<.0001 and RR=9.3, p<.0001], life-threatening thrombosis during evolution [RR=35.4, p<.0001 and 12.8, p<.0001], infections during evolution [RR=6.6, p=.015 and RR=1.5, p=.17], need for subsequent therapy [RR=5.4, p=.041 and RR=3.0, p=.0001] and age over 54 years [RR=4.39, p=.048 and RR=4.16, p<.0001]. Furthermore, in *de novo* PNH patients, MDS or AL were related to poor survival [RR=14.3, p<.0001] (no malignancy in patients with the AA-PNH syndrome), as was the occurrence of pancytopenia during evolution [RR=4.4, p<.0001] (not applicable to patients with the AA-PNH syndrome). Finally, in the univariate analysis, patients with the AA-PNH syndrome were shown to have a survival advantage over patients with *de novo* PNH [RR=0.33, p=.007] (**Figure 2**).

Table 4. Multivariate analysis of factors influencing survival.

Factor	Grafted patients censored		
	RR	95% CI	P
Risk factor with similar impact in AA-PNH & *de novo* PNH			
MDS or AL (evolution)*	20.5	6.6-63.6	<.0001
Pancytopenia (evolution)*	8.4	3.8-18.8	<.0001
Age > 54 years*	3.6	2.0-6.3	<.0001
Thrombocytopenia at diagnosis	2.4	1.3-4.4	.004
Thrombosis (evolution)*	14.5	7.9-26.4	<.0001
Impact of treatment in the outcome in AA-PNH & *de novo* PNH			
No Androgens as first line treatment	5.12	2.0-13.8	.0001
No Androgens as first line treatment x AA-PNH syndrome	.045	.005-.44	.0001

* time-dependent covariate. RR; relative risk mean estimate. 95% CI; relative risk 95% confidence interval.

In multivariate analysis done after censoring patients who had undergone bone marrow transplantation the following factors were related to poor survival: development (as a complication) of thrombosis, MDS or AL, progression to pancytopenia in patients with *de novo* PNH, age over 54 years, thrombocytopenia at diagnosis and initial treatment without androgens. Specifically, there was a strong interaction between initial treatment without androgens and the AA-PNH syndrome [RR=0.44, p=.0001]. A second but less significant interaction was found between the occurrence of thrombosis and the AA-PNH syndrome [RR=4.5, p=.043]. The results of the multivariate analysis done with censoring patients at the time of transplantation are given in **Table 4**. The same factors and interactions, with slightly different RRs, were related to survival. In this second model, the AA-PNH syndrome was no longer related to better survival (p>0.29). No deviation to the Cox' proportionality assumption could be detected in both final models.

Discussion

Since the seminal description of its symptoms by Paul Strübing in 1882 [5], PNH has captured the attention of physicians and researchers. PNH is an acquired chronic hemolytic anemia, often associated with recurrent nocturnal exacerbation, neutropenia and/or thrombocytopenia, and episodes of venous thrombosis. The biological hallmark of PNH cells is the deficiency in cell surface proteins that share the common structural feature of being anchored to the cell surface by a glycosyl-phosphatidylinositol moiety (GPI) [6]. Recently, the *PIG-A* gene was identified using expression cloning [7-8]; its somatic molecular alterations were shown to be responsible for the defect in GPI-anchored protein expression in all

PNH patients studied today [9], including patients with the AA-PNH syndrome [10].

From the clinical point of view, the interrelationship between AA and PNH has long been recognized. More than 50 years ago, Dacie and Gilpin described the first patient with an AA-PNH syndrome [11] and in 1967 Dr Dameshek stated in an editorial that " *Aplastic anemia, and PNH ... might have a common denominator in the form of an insult to the marrow*" [12]. Dacie & Lewis [4] have already stated that these two diseases are associated in two ways:

(1) A patient is diagnosed with typical PNH and later develops pancytopenia. This feature was thought to be uncommon [1, 13], but in fact this represents, in our experience, an evolution of the disease in 23 out of 155 patients with *de novo* PNH (15%) and the cumulative incidence of this particular complication is 22% at 10 years [2]. Furthermore, as in our first analysis, these patients with *de novo* PNH developing pancytopenia were shown to have a dismal prognosis with a relative risk of death of 8.4, adjusted for all other prognostic factors.

(2) Another patient is diagnosed with AA, and the Ham's test becomes positive later. This patient subset presenting with a so-called AA-PNH syndrome represents 30% of our patients, a figure similar to that reported by Dacie & Lewis (23 out of 80 cases) [4] but less than that reported recently by the Ulm and Japanese groups [14-15]. This apparent discrepancy may in fact only reflect the bias of referral to centers and/or random fluctuations in relation to sample size. On the other hand, in series involving large numbers of patients with AA, an AA-PNH syndrome is increasingly recognized as a late complication occurring in patients who have been treated with immunosuppressive therapy [16]. Although the exact incidence of PNH in patients with AA is not clearly known (most probably due to changes in basic knowledge and increased recognition of this disease entity), a conservative estimate from large series is that between 15%-25% of AA patients will eventually develop a PNH clone [see the chapter by G Socié & A Tichelli, this volume].

In a first study, we found, in a multivariate analysis, that patients with the AA-PNH syndrome have a better prognosis (3 times less risk of death) compared to patients with *de novo* PNH. All these patients were previously treated [2]. We had no evidence that they received a different therapy or transfusion support, from the standard treatments for AA during the given period of time, and we had no clear explanation of this finding of a better survival for the patients with the AA-PNH syndrome. We did not know if the AA-PNH syndrome was a different disease entity as compared to the *de novo* form of the disease, but we thought that further analysis of this patient cohort might be able to help solve the question. In fact, to the best of our knowledge, none of the large series of patients with PNH nor those of patients with AA affords direct comparisons between patients with *de novo* PNH and those with the AA-PNH syndrome. The present study, on a large group of patients with this disease, allowed for such a comparison. At diagnosis, presenting symptoms were similar although patients with *de novo* PNH tended to be more frequently recognized through the association of abnormal hematological values and abdominal pain crises. Blood examination revealed that patients with the AA-PNH syndrome were more frequently pancytopenic than patients with *de novo* PNH (although not statistically significant). However, when comparing patients with the AA-PNH syndrome *vs. de novo* PNH at presentation, anemia was less frequently observed, while thrombocytopenia and neutropenia were more frequently detected. These blood abnormalities fit with those reported in *de novo* PNH and in AA patients in "clinical remission" after immunosuppressive therapy. Treatment options showed variations with time but also clearly reflected physician options with regard to both groups, since patients with the AA-PNH syndrome were more frequently treated with immunosuppressive therapy or with androgens, than patients with *de novo* PNH. Perhaps surprisingly, during follow-up, the prevalence of infections or abdominal pain crises, cumulative incidence of MDS or acute leukemia and the cumulative proportion of patients requiring additional treatment were similar in both groups. Finally, an analysis of primary causes of death revealed some differences between the two groups. This was particularly true for CNS thrombosis and the Budd-Chiari syndrome.

In the last part of this study, we searched for factors related to survival through stepwise multivariate Cox models and tested, through interaction terms, whether the relationship of each factor to survival was different in the two groups. The results of the analysis with grafted patients censored at the time of transplantation first confirmed, and refined, our previous analysis [2]: 4 factors influenced survival in both groups (life-threatening thrombosis, MDS or AL, age over 54 years and thrombocytopenia at diagnosis), while one factor (pancytopenia during the evolution) was a risk factor in patients with *de novo* PNH only (per design of the study). But what this first analysis helps to clarify is why patients with the AA-PNH syndrome had an apparently better prognosis. This was solved by introducing interaction terms into the model to reflect the major impact of the primary therapy in both groups of patients. In fact, treatment other than androgens led to an increased risk of death (~ 5 fold) in patients with *de novo* PNH, while it lowered the risk of death (~ 4 fold) in patients with the AA-PNH syndrome, relative to patients treated by androgens in either group. In the final model, including this interaction, there was no longer any apparent survival advantage for patients having an AA-PNH syndrome.

References

1. Rosse, W.F. Paroxysmal nocturnal hemoglobinuria. In Principles and practice of hematology; Stossel TP & Lux SE, eds (1995): Lippincott, Philadelphia: chapter 11; pp 367-76
2. Socié, G., Mary, J.Y., De Gramont, A., *et al* (1996). Paroxysmal nocturnal hemoglobinuria: long-term follow up and prognostic factors. *Lancet* **348:** 573-577.
3. Lewis, S.M. & Dacie J.V (1967). The aplastic anaemia- paroxysmal nocturnal hemoglobinuria syndrome. *Br J Haematol.* **13:** 236-251.

4. Dacie, J.V. & Lewis, M (1972). Paroxysmal nocturnal haemoglobinuria: clinical manifestation, haematology, and nature of the disease. *Ser Haematol.* **3:** 3-23.
5. Strübing, P (1882). Paroxysmale Haemoglobinurie. *Deutsche Med Woch,* **8:** 1-3.
6. Yeh E.T.H., Rosse, W.F (1994). Paroxysmal nocturnal hemoglobinuria and the glycosylphosphatidylinositol anchor. *J Clin Inv* **93:** 2305-2312.
7. Bessler, M., Mason, P. J., Hillmen, P., *et al.*(1994) Paroxysmal nocturnal haemoglobinuria (PNH) is caused by somatic mutations in the PIG-A gene. *EMBO J* **13:** 110-117.
8. Takeda, J., Miyata, T., Kawagoe, *et al.* (1993) Deficiency of the GPI anchor caused by a somatic mutation of the *PIG-A* gene in paroxysmal nocturnal hemoglobinuria. *Cell* **73:** 703-711.
9. Luzzatto, L. & Bessler, M. (1997) The dual pathogenesis of paroxysmal nocturnal hemoglobinuria. *Current Op Hematol* **3:** 101-110.
10. Nagarajan, S., Brodsky, R.A., Young N.S., Medof M. E. (1995) Genetic defects underlying paroxysmal nocturnal hemoglobinuria that arises out of aplastic anemia. *Blood* **86:** 4656-4661.
11. Dacie, J.V. & Gilpin A. (1944) Refractory anaemia (Fanconi type) : its incidence in three members of one family, with in one case a relationship to chronic haemolytic anaemia with nocturnal haemoglobinuria. *Arch Dis Child* **19:** 155-162.
12. Dameshek, W. (1967) What do aplastic anemia, paroxysmal nocturnal hemoglobinuria and "hypoplastic" leukemia have in common ? *Blood* **30:** 251-255.
13 Hillmen, P., Lewis, S.M., Bessler, M., Luzzatto, L., Dacie, J.V. (1995) Natural history of paroxysmal nocturnal hemoglobinuria. *New Engl J Med* **333:** 1253-1258.
14. Späth-Schwalbe, E., Schrezenmeier, H., Heimpel, H. (1995) Paroxysmale nätliche hämoglobinurie: Klinische erfahrungen bei 40 patienten in einem zentrum über 25 jahre. *Deutsche Med Woch* **120:** 1027-1033.
15. Fujioka, S. & Takayoshi, T. (1989) Prognostic features of paroxysmal nocturnal hemoglobinuria in Japan. *Acta Haematol Jap* **2:** 1386-1394.
16. Tichelli, A., Gratwohl, A., Nissen, C., Speck, B. (1994) Late clonal complications in severe aplastic anemia. *Leuk Lym* **12:** 167-175.

Part 4

Mechanism of Clonal Expansion in PNH

PNH clone acquires both a survival and a growth advantage ?

Hideki Nakakuma and Tatsuya Kawaguchi

Second Department of Internal Medicine, Kumamoto University School of Medicine, Honjo 1-1-1, Kumamoto 860-8556, Japan

Summary: The cloning of the *PIG-A* gene has facilitated the unraveling of complex pathophysiology in paroxysmal nocturnal hemoglobinuria (PNH). Of current major interest is the mechanism by which PNH clone expands to manifest clinical symptoms. Gene targeting showed that clonal dominance is not attributable to *PIG-A* mutation alone. For the expansion, there are two hypotheses of a survival and a growth advantage. Increasing evidence has shown that an immune mechanism injures blood cells in patients with bone marrow failure syndromes (BFS) such as PNH, aplastic anemia and myelodysplastic syndromes. The appearance of cells with *PIG-A* mutation in patients with BFS suggests their escape from the immune attack. In fact, we found that blood cells with *PIG-A* mutation were less susceptible than the control counterpart with normal phenotype to natural killer (NK) cells in vitro. We then realized a survival advantage, although the pathological role of NK cells is not established yet in BFS. On the other hand, we have often observed a clone expanding distinctly among multiple clones with the same PNH phenotype in a single patient. It is likely then that factors other than *PIG-A* mutation and immune attack participate in the clonal expansion. Indeed, we have detected a preferential hematopoiesis by PNH stem cells when transplanted in mice, supporting the intrinsic growth alteration. Thus, we conclude that PNH clone acquires both a growth and a survival advantage. We have further detected increased mutations of the hypoxanthine-guanine phosphoribosyltransferase gene and proposed mutable conditions that favor the emergence of diverse mutant clones with various advantages in PNH.

Key words: PNH, *PIG-A*, NK cells, *HPRT*.

PIG-A mutation confers a relative survival advantage to PNH clone in cytotoxic attack by NK cells

We have tried to verify a survival advantage theory for the pathogenic expansion of PNH clone [1,2]. A human myeloid leukemia clone (K562) with *PIG-A* mutation was transfected with a vector either containing or deficient in *PIG-A* cDNA [3,4]. The *PIG-A* transfectant restored the membrane expression of glycosylphosphatidylinositol (GPI)-linked proteins such as decay-accelerating factor (DAF) and CD59 (revertant, a control counterpart), whereas the mock-transfectant still failed to express the proteins (*PIG-A* mutant). There was no difference in cell growth between the *PIG-A* mutant and revertant clones. The pair were then used as target cells in a cytotoxicity assay with natural killer (NK) cells which is a candidate for the effector lymphocytes that exert injury on hematopoietic cells in patients with bone marrow failure syndromes (BFS) including PNH, aplastic anemia and myelodysplastic syndromes (MDS) [5]. In a cytotoxicity assay with ^{51}Cr-release, the *PIG-A* mutant was less susceptible to NK cells than the control counterpart (Fig. 1). Further experiments, using a cultured NK cell line (KHYG-1) [6] instead of peripheral NK cells or using B cell line (JY5) [4] instead of K562, also showed the decreased sensitivity of *PIG-A* mutants to NK cells. Thus, these results suggest that *PIG-A* mutations decrease the susceptibility to NK cells.

We further investigated the mechanism by which NK cells killed the target cells. Because the cultured NK cells that lack Fas ligand [6] showed the cytotoxicity, we considered that the NK cells used other pathways than Fas/Fas ligand to kill the target cells. We then examined the effects of concanamycin A (CMA) and a calcium chelate EGTA that inhibit the perforin/granzyme pathway [7,8]. Each reagent inhibited dose-dependently and eventually abolished the killing of both mutant and control cells. These findings suggest that NK cells used the perforin/granzyme pathway alone. The mutant and control cells showed the same sensitivity to purified perforin. It is then conceivable that the decrease in the sensitivity of *PIG-A* mutant cells to NK cells reflects an impaired attachment between NK cells and the mutant cells. Of note, none of the antibodies to DAF and CD59 showed any effect on the cytotoxicity. GPI-linked membrane molecules responsible for the decreased susceptibility of *PIG-A* mutant cells to NK cells remain to be identified.

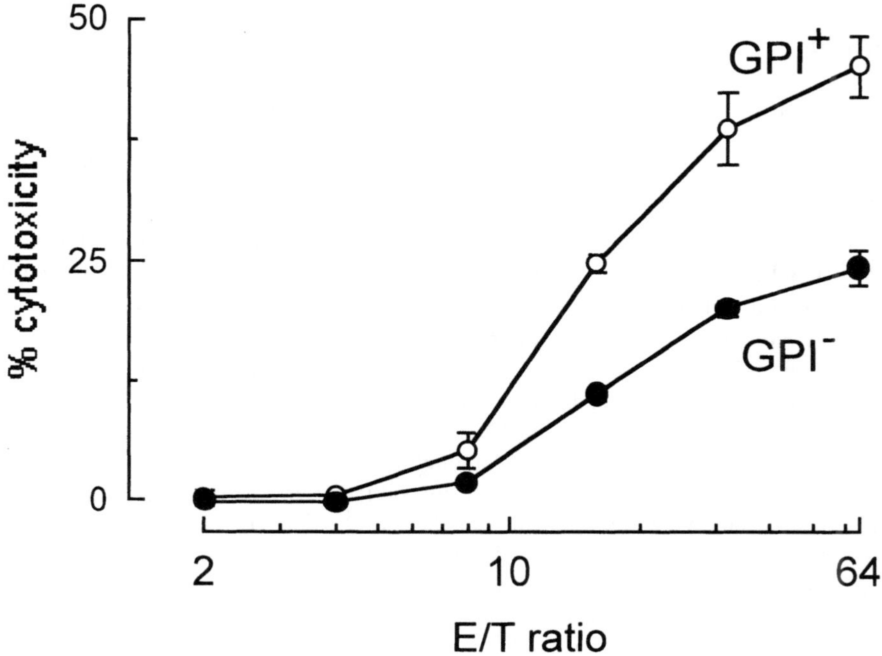

Fig. 1. Cytotoxicity assay. NK cells isolated from a healthy donor were incubated with ^{51}Cr-labeled target cells (either GPI^+ control or GPI^- mutant K562 cells) for 3 hours. Cytotoxicity was assessed by the specific ^{51}Cr release. Data are shown as the mean $\pm$ SD of triplicate assays. The killing of GPI^- cells were less than that of GPI^+ cells.

Preferential hematopoiesis by PNH stem cells in SCID mice

To know whether PNH clone harbors some neoplastic features comprising an intrinsic growth abnormality, we characterized the in vivo growth by the analysis of hematopoietic reconstitution with PNH stem cells transplanted into sublethally irradiated severe combined immunodeficient (SCID) mice [9]. In the mice, bone marrow $CD34^+$ cells of patients with PNH generated blood cells of 3 lineages with PNH phenotype alone (Fig. 2). The hematopoiesis persisted for more than 6 months and did not require any human cytokines. In contrast, $CD34^+$ cells of healthy volunteers needed an intense stimulation with hematopoietic cytokines (interleukin-3, stem cell factor, granulocyte-macrophage colony-stimulating factor) to generate human blood cells in mice. From this cytokine-independent preferential hematopoiesis by the PNH stem cells, it is likely that the PNH clone had an intrinsic growth alteration. Although the immune system of the SCID mice was impaired genetically and further by sublethal irradiation, NK cells might reside and survey immunologically in the SCID mice. In this regard, NK cells decreased in the mice and barely killed human blood cells in vitro. Thus, these findings indicate an intrinsic growth alteration rather than an escape from NK cells. The possibility that PNH clone exhibits both a survival and a growth advantage is not excluded.

Increased mutations of the hypoxanthine-guanine phosphoribosyltransferase gene (*HPRT*)

Coexistence of clones with different *PIG-A* mutations [10-12] and clones of MDS and leukemia in the same PNH patient [13,14] suggests that mutations are not limited to specific clones or genes, rather, they may be random events. To know whether these events manifest the conditions that facilitate somatic mutations, we determined the mutational frequency of the *HPRT* gene which reflects overall genetic mutation rates [15] and characterized the nature of the mutations in T cells. Using a methylcellulose culture, we detected *HPRT* mutant clones that formed colonies in the presence of 6-thioguanine. Results showed that the *HPRT* gene mutated more frequently in PNH than healthy controls. The mutation spectrum consisted primarily of base deletions and was similar to that of *PIG-A* in published literature [16], indicating that both genes were exposed to common mutagens. In contrast, healthy controls showed only single base substitutions. Thus, we conclude that in PNH patients, conditions exist that favor the occurrence of diverse somatic mutations besides spontaneous mutations.

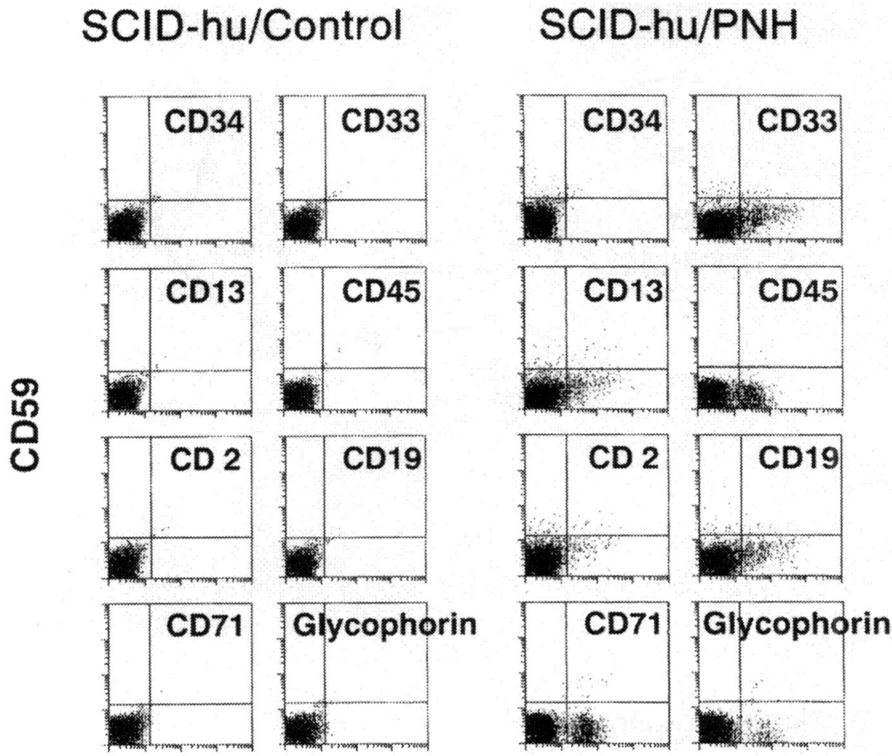

Fig. 2. Flow cytometric analysis of human cells found in bone marrow of SCID mice 7 months after transplantation with CD34$^+$ cells. SCID-hu/Control (left panel) and SCID-hu/PNH (right panel) represent the mice transplanted with cells of a healthy donor and a patient with PNH, respectively. Human blood cells were detected in only SCID-hu/PNH and all negative for CD59.

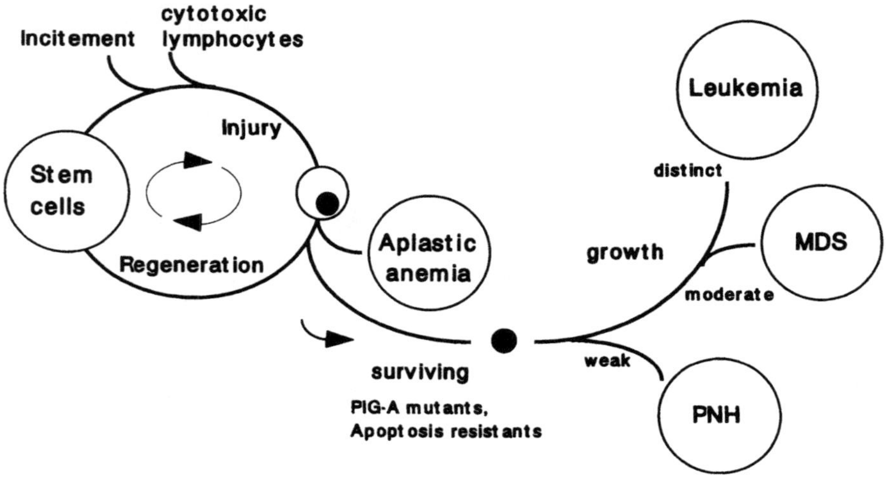

Fig. 3. A model for pathogenesis of PNH and related bone marrow failure syndromes.

Working hypothesis

Fig 3 shows a model for the pathogenesis of PNH and related diseases. In the presence of incitement (virus infection is a plausible candidate), hematopoietic stem cells become targets of cytotoxic lymphocytes (NK and T cells), resulting in marrow hypoplasia. In the marrow, cells that are less sensitive to the injury survive and are concentrated. *PIG-A* mutant is one of those with this survival advantage. When surviving cells further acquire various degrees of proliferative advantage by additional mutations, they expand to develop PNH, MDS and leukemia.

References

1. Young NS, Maciejewski J (1997) The pathophysiology of acquired aplastic anemia. New Engl J Med 336:1365-1372
2. Luzzatto L (1999) Paroxysmal murine hemoglobinuria (?): a model for human PNH. Blood 94:2941-2944
3. Hirose S, Mohney RP, Mutka SC, et al (1992) Derivation and characterization of glycoinositol-phospholipid anchor-defective human K562 cell clones. J Biol Chem 267:5272-5278
4. Nishimura J, Phillips KL, Ware RE, et al (2001) Efficient retrovirus-mediated PIG-A gene transfer and stable restoration of GPI-anchored protein expression in cells with the PNH phenotype. Blood 97:3004-3010
5. Dunn DE, Tanawattanacharoen P, Boccuni P, et al (1999) Paroxysmal nocturnal hemoglobinuria cells in patients with bone marrow failure syndromes. Ann Intern Med 131:401-408
6. Yagita M, Huang CL, Umehara H, et al (2000) A novel natural killer cell line (KHYG-1) from a patient with aggressive natural killer cell leukemia carrying a p53 point mutation. Leukemia 14:922-930
7. Kataoka T, Shinohara N, Takayama H, et al (1996) Concanamycin A, a powerful tool for characterization and estimation of contribution of perforin- and Fas-based lytic pathways in cell-mediated cytotoxicity. J Immunol 156:3678-3686
8. Stenger S, Rosat J, Bloom BR, et al (1999) Granulysin: a lethal weapon of cytotoxic T cells. Immunol Today 20:390-394
9. Iwamoto N, Kawaguchi T, Horikawa K, et al (1996) Preferential hematopoiesis by paroxysmal nocturnal hemoglobinuria clone engrafted in SCID mice. Blood 87:4944-4948
10. Bessler M, Mason P, Hillman P, et al (1994) Somatic mutations and cellular selection in paroxysmal nocturnal haemoglobinuria. Lancet 343:951-953
11. Endo M, Ware RE, Vreek TM, et al (1996) Molecular basis of the heterogeneity of expression of glycosylphosphatidylinositol anchored proteins in paroxysmal nocturnal hemoglobinuria. Blood 87:2546-12557
12. Nishimura J, Inoue N, Wada H, et al (1997) A patient with paroxysmal nocturnal hemoglobinuria bearing four independent PIG-A mutant clones. Blood 89:3470-3476
13. Ishihara S, Nakakuma H, Kawaguchi T, et al (2000) Two cases showing clonal progression with full evolution from aplastic anemia-paroxysmal nocturnal hemoglobinuria syndrome to myelodysplastic syndromes and leukemia. Int J Hematol 72:206-209
14. Iwanaga M, Furukawa K, Amenomori T, et al (1998) Paroxysmal nocturnal haemoglobinuria clones in patients with myelodysplastic syndromes. Br J Haematol 102:465-474
15. Albertini RJ, Nicklas JA, O'Neill JP, et al (1990) In vivo somatic mutations in humans: measurement and analysis. Annu Rev Genet 24:305-326
16. Kinoshita T, Inoue N, Takeda J (1995) Defective glycosyl phosphatidylinositol anchor synthesis and paroxysmal nocturnal hemoglobinuria. Adv Immunol 60:57-103

THE ROLE OF LYMPHOID CELLS IN THE PATHOGENESIS OF PNH

Lucio Luzzatto[1,2], Anastasios Karadimitris[2,3], David Araten[2], Rosario Notaro[1,2]

[1] Istituto Nazionale per la Ricerca sul Cancro, Genova, Italy;
[2]Department of Human Genetics, Memorial Sloan-Kettering Cancer Center, New York, NY, USA;
[3]Dept of Haematology, Hammersmith Hospital, Imperial College, London, UK

Abstract: Paroxysmal Nocturnal Hemoglobinuria (PNH) is an acquired clonal disorder of the hematopoietic stem cell (HSC), characterized by blood cells deficient in the glycosylphosphatidylinositol (GPI)-anchored proteins. The GPI-deficient blood cells result from a somatic mutation of the X-linked *PIG-A* gene in one or more HSC of the PNH patients, in whom the HSC carrying the mutated *PIG-A* gene expand; to the extent that in many cases GPI-deficient cells eventually comprise most of hematopoiesis. In PNH patients the expansion of the GPI-deficient hematopoiesis is almost invariably associated with features of bone marrow failure. This fact, together with other clinical and experimental evidence, suggests that the expansion of *PIG-A* mutant clones is due to a concomitant selective process against the normal HSC carrying the wild type *PIG-A*, rather than to an intrinsic growth advantage of the mutated HSC.

The close association between PNH and Idiopathic Aplastic Anaemia, and the recent finding of increased frequency of expanded T cell clones in patients with PNH, support a pathogenetic model based on autoreactive T cells that cause the selective depletion of normal HSC, whereas PNH HSC escape this T-cell-mediated attack, survive and expand. However, the mechanism responsible for the bone marrow failure and the target of the autoaggressive T cells remains still unknown. Here, we review the the most recent findings and possible mechanisms.

Keywords: PNH, T cells, cytotoxicity, T-LGL disease, PNH and non-PNH targets.

PNH has been traditionally and appropriately classified as a 'red cell disorder', because chronic hemolytic anemia with intravascular hemolysis is certainly its most spectacular and characteristic manifestation [1,2]. However, hints that there was more to the pathophysiology of PNH than hemolysis are anything but new, since the development of pancytopenia in patients with PNH was reported since the fifties [3]; and by 1972 Dacie and Lewis published a classical review on the

"PNH/Aplastic Anemia (AA) syndrome"[4]. By the late eighties it was possible to ask the question, whether there was a continuum between PNH and AA, or whether in fact these were different stages or different aspects of the same pathophysiological entity [5]. We suggested at the time that the latter was the case [6], and that there are *always* two factors contributing to, indeed required for the pathogenesis of PNH: (a) an abnormal blood cell population, with exquisite hypersusceptibility to complement, arising from a somatic mutation; (b) an element of bone marrow failure (BMF). Depending on the relative severity of these two components, and depending on the time relationship in their development, a patient would present with the clinical picture of PNH, or of AA, or of the PNH/AA syndrome.

It is appropriate that this meeting is being held in Japan, because the cloning of the *PIG-A* gene, achieved by Taroh Kinoshita's group in Osaka, has been a milestone in the history of PNH [7], as it has essentially solved the molecular basis of component (a): indeed, mutational inactivation of *PIG-A*, characteristic of PNH, causes severe deficiency of all glycosylphosphatidylinositol (GPI)-linked proteins[8], including CD59 and CD55, the main protectors of red cells against activated complement [9]. Hence, the outstanding issue in the pathophysiology of PNH is (b); or, more exactly, the relationship between BMF and the PNH cell population [10,11].

The basis for the expansion of a PNH cell population

Specific *PIG-A* mutations tell us that in each patient the PNH cells belong to one or to a few clones: nevertheless, PNH cells constitute a substantial percentage(s) [12]. The possibility that the expansion of PNH clones is just relative, rather than absolute, can be quickly dismissed, because in some patients a PNH clone may – on its own – support the patient's entire hematopoiesis with normal or near normal blood counts (see Fig. 1). Therefore, we must explain a massive clonal expansion in absolute terms. There are essentially two ways in which this can take place. (i) The expansion in absolute terms of a PNH clone is due to an intrinsic growth advantage. (ii) The expansion of the PNH clone is absolute, but the growth advantage is only relative, by comparison to the growth of normal hematopoietic cells in a certain environment: *i.e.*, we are dealing with a *conditional* growth advantage.

At the moment, there are several lines of evidence favoring the latter possibility (see Table 1). Perhaps the most compelling of all is that small PNH clones exist in all normal people [13-15], and that the very same *PIG-A* mutation has been found in normal people and in PNH patients [13] (see Table 2). One might argue that perhaps the PNH clone in a PNH patient has an additional mutation in an unknown gene, and this remains a formal as yet untested possibility. However, the follow up of patients with PNH generally reveals a remarkable stability of the

PNH cell population, as though the clonal expansion has reached a certain size, but then proceeds no further [16].

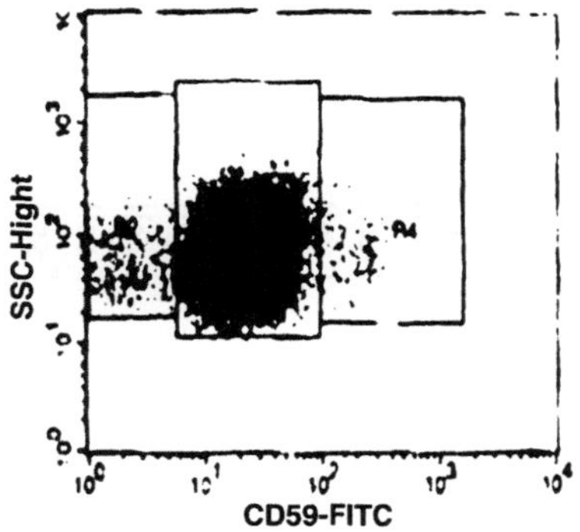

Fig.1. Flow cytometry analysis of red cells from a 34 year-old man with a 9-year history of PNH characterized by multiple thrombotic events. It is seen that about 98% of the patient's red cells are PNH II, with only about 1% normal red cells and about 1% PNH III red cells. At the time of the study the patient's blood count revealed a Hb level of 11.7 G/dl, WBC 7400, platelets 192,000. Thus, in this patient at this time a virtually normal hematopoiesis was provided by a single PNH clone and therefore by a single stem cell.

Table 1. MECHANISM OF EXPANSION OF PNH CLONES

Lines of evidence	*In favour of intrinsic growth advantage of PNH clone*	*In favor of growth advantage of PNH clone being only conditional*
Clinical	Occasionally PNH evolves to acute myeloid leukemia (AML) [25-28]	In most cases the size of PNH cell populations is stable [16]
Experimental	PNH cells may be more resistant to apoptosis[31] (controversial [32])	Resistance to apoptosis may be dependent on level of *fas* ligand [33]. In mouse models PNH cell populations tend to contract and not to expand [34-36]
Molecular		PNH clones are present in normal people (in the absence of selection) [13-15] Several expanded clones may co-exist in same PNH patient at the same time (convergent selection) [37,38] Different PNH clones may account for PNH in same patient at different times (persistent selection) [39]

Table 2. PIG-A mutations in normal individuals

Patient ID#	Exon	Mutation*		Amino acid replacement	cell source	Reference
3	6	1381	C>T	461 Arg > Trp	PMN	[13]
4	2	196	Ins AT	frameshift (stop 68)	PMN	"
5	2	614	T>A	205 Val > Asp	PMN	"
8	2	229	C>T[$]	77 Arg > stop	PMN	"
9	2	167	T>C	56 Leu > Pro	PMN	"
10	2	294	C>A[$]	98 Tyr > stop	PMN	"
5	2	76	ΔT	frameshift (stop 60)	T-cell	[14]
2	2	549	T>A	183 Cys > Stop	T-cell	[15]
	5	Δ1018–32		Δ340–344	T-cell	"
3	5	Δ1013–32		frameshift (stop 342)	T-cell	"
	2	349	Δ T	frameshift (stop 124)	T-cell	"
4	2	189	Δ A	frameshift (stop 68)	T-cell	"
5	2	389	Δ C	frameshift (stop 171)	T-cell	"
	2	548	G>A	183 Cys > Tyr	T-cell	"
6	3	729	Δ T	frameshift (stop 247)	T-cell	"
	2	Δ361–3	Δ 121		T-cell	"
9	2	383	A>G	128 His > Arg	T-cell	"
12	2	394	Δ T	frameshift (stop 171)	T-cell	"

Patient ID according to the original report
*Numbers indicate position in the cDNA sequence starting with the first coding nucleotide
$ The same mutation has been previously reported in a PNH patient

Mechanism of conditional growth advantage of PNH cells

Thus, the weight of the evidence currently available favors the view that PNH cells have a conditional growth advantage. On the other hand, we have already noted that there is always an element of BMF in PNH: therefore, it is reasonable to consider that these two facts may not be unrelated. Indeed, it is tempting to surmise that the same process damages normal hematopoietic cells and spares PNH hematopoietic cells [17], thus favoring their expansion. According to this model [6], we are dealing in formal terms with negative selection against normal hematopoietic cells (HSC), rather than positive selection for PNH cells.

The nature of the selective agent remains at the moment hypothetical: however, in the case of aplastic anemia there is evidence that BMF may be the consequence of an auto-immune process [18]. In principle, an auto-immune process might damage selectively normal HSCs more than PNH HSCs because the former have an intrinsically greater sensitivity to cytotoxicity. We have carried out experiments in order to test specifically this possibility, by making use of lymphoblastoid cell lines (LCL) which we have isolated from PNH patients by transformation with Epstein-Barr virus (EBV) of peripheral blood B cells. By using limiting dilutions we have then isolated clones, some of which are GPI(+) and others are GPI(-). We have exposed T cells from PNH patients to EBV- transformed GPI(+) B-LCLs previously isolated from the same patient: thus, the main antigen driving T-cell proliferation was EBV, and the entire system was autologous. After several rounds of growth, the T cells were tested against GPI(+) B-LCLs and GPI(-) B-LCLs in cytotoxicity assays: no significant differences could be detected[19]. The same was true with EBV-specific T-cell clones obtained from the bulk cultures. Thus, at least by this approach we could find no evidence for a deceased vulnerability of PNH cells at the effector stage of a potent T-cell mediated immune reaction. Very recently, however, Nakakuma et al have reported that the rate of killing by NK cells of myeloid and lymphoid PNH-like cell lines was significantly less than that of non-PNH-like cells.[20]

Unlike other mechanisms of toxicity, an auto immune process is potentially endowed with a virtually unlimited degree of target specificity. Therefore, auto reactive T cells are in principle excellent candidates for being able to display subtlety of discrimination, as required to recognize normal HSC from PNH HSCs. In fact, we already know that all GPI-linked surface molecules are deficient in the latter: therefore, at first sight each one of these is a potential target for an auto immune attack which would damage normal HSCs but would leave PNH HSCs unscathed.

Search for auto reactive T cells in PNH patients

A sensitive way to explore the T-cell repertoire is the so-called immunoscope analysis [21]. We have carried out on peripheral blood mononuclear cells a systematic analysis of the size distribution of the complementarity determining region 3 (CDR3) of the variable region of the β chain (βV) of the TCR by cDNA amplification, in 24 parallel PCR reactions, with 24 TCR βV family-specific 'sense' primers and a common TCR βC antisense primer [21]. By comparing 19 PNH patients, 15 age-matched normal individuals, and 10 multiply transfused patients as an additional control group, we found (see Fig. 2) that PNH patients had a more than three fold increase in the frequency of skewed patterns within the 24 TCR βV families (30%±16.2 v 8.9%±10.4, P=0.01). When individual βV families were compared, the frequency of skewing was also significantly higher in PNH patients in several cases. By using βJ segment-specific primers and, in selected cases, direct nucleotide sequencing, we found that the skewed TCR βV patterns were due to the presence of T-cell clones of significant size. In a longitudinal study we found that in two patients who had not received any type of immunosuppressive treatment, the TCR βV CDR3 profiles remained unchanged over a period of 9 and 19 months respectively. By contrast, in one patient who had a good response to immunosuppressive treatment with anti-thymocyte globulin (ATG) the pre-ATG TCR βV CDR3 profile abnormalities had reverted to near normal. In another patient, whose hematological parameters were essentially unchanged after ATG treatment, the pre- and post-ATG TCR βV CDR3 profile abnormalities remained almost identical [22].

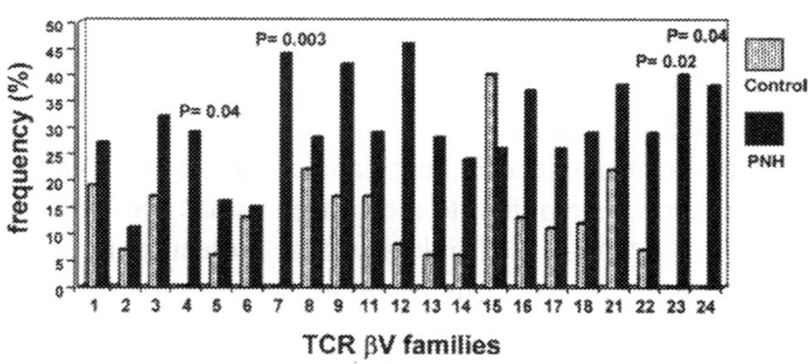

Fig. 2. Increased skewness of T-cell repertoire in several TCR β families in patients with PNH (modified from [22]).

Association between expanded T-cell clones and PNH

Recently we have observed another link between T-cell abnormalities and PNH. In 3 patients with classical PNH we have found expanded T-cell clones with the morphology and the immunophenotype of large granular lymphocytes (LGL): one of them (Fig. 3) meets all criteria for LGL leukemia. In this the LGLs were characterized extensively: they were CD8+, CD57+, CD59+: thus, they had a 'NKT" phenotype. Their clonal nature was proven by the finding of a discrete rearranged band on a Southern blot hybridised to a β-TCR probe. We have observed that the large T-LGL population emerged from a background of multiple smaller T-cell clones. The identity of the TCR gene was determined, and the rearranged TCR β chain was partially sequenced [23].

PNH and T-LGL are both so rare, that we can safely discount the possibility that they co-existed in the same patient by chance alone. It does not seem likely that PNH caused T-LGL, especially since the LGL population did not belong to the PNH clone; we believe it is much more likely that LGLs are causative in the development of PNH. In fact, this patient, like many other PNH patients, had a markedly skewed T-cell repertoire. We suggest that in general auto-reactive T-cell clones in PNH patients expand initially as a result of antigenic stimulation, and then play a major role in the suppression of non-PNH hematopoiesis. The same thing may have been true in this patient, who probably did not have a neoplastic T-cell disorder *ab initio*; rather, we speculate that her expanded T-cell population increased the target size for a genetic event to cause neoplastic transformation.

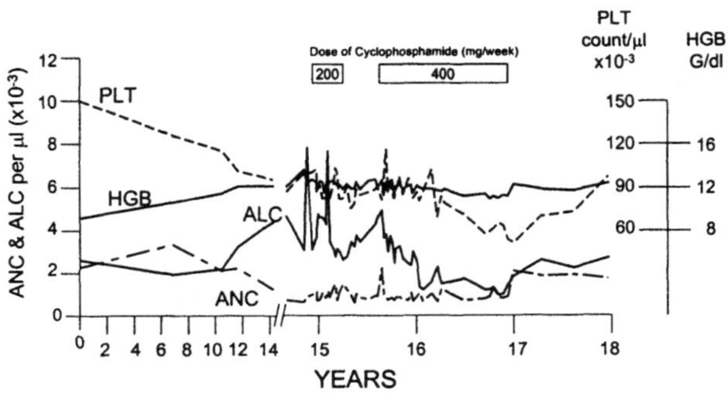

Fig. 3. Clinical course of a patient with PNH and large granular lymphocyte leukemia (modified from [40]).

Search for a T-cell target absent on PNH cells

Since in PNH, as in AA, there is evidence for an auto immune process, and since we already have evidence for abnormalities in the T-cell repertoire, a specific question at the moment is to identify the target of auto reactive T cells. At the moment, we can consider at least three possibilities. (i) *The target of autoaggressive T cells is an epitope belonging to the PIG-A protein itself.* This is not very likely for at least two reasons. First, since PIG-A is ubiquituosly expressed we would expect an autoimmune disease not limited to hematopoiesis. Second, with missense mutations in *PIG-A* PNH cells would remain vulnerable to a PIG-A dependent attack. (ii) *The target of autoaggressive T cells is an epitope derived from a GPI-anchored protein.* Proteins destined to be GPI-linked fail to land on the surface of PNH cells because they do not find the anchor: however, we know that the respective genes efficiently transcribed in PNH cells, and the transcripts are probably translated. This results, in PNH cells, in the accumulation of unprocessed proteins which, in association with ER-resident chaperones, are transferred to the cytoplasm where they are degraded. It is not yet clear to what extent this may favor or limit presentation of resulting peptides by class I and/or class II MHC molecules. Thus, any of these proteins remains a potential target.

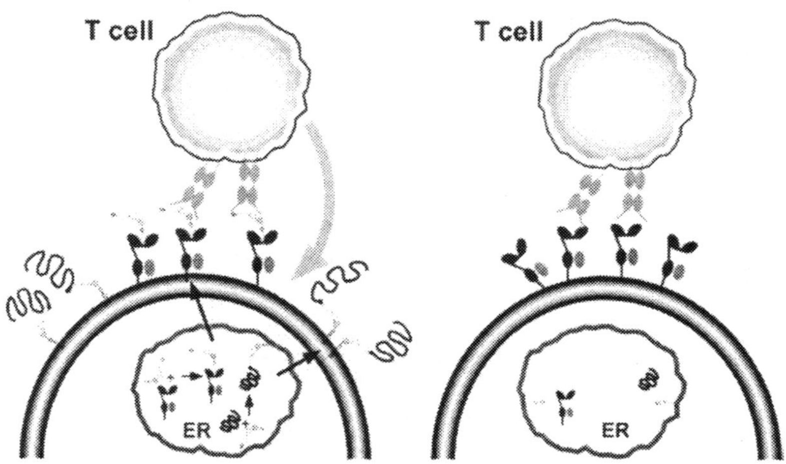

Fig. 4. The GPI anchor is a potential target of auto reactive T cells in PNH (from [41]). In this model GPI-specific, CD1d-restricted T cells, target GPI+ HSCs (left). By contrast, GPI- HSCs are spared, because they fail to synthesise the GPI anchor (right). As a result, there is a depletion of normal stem cells and an expansion of the PNH cell population, which can increasingly contribute to hematopoiesis (modified from [41]).

(iii) *GPI itself is the target of autoaggressive T cells.* It is now well established that, just as HLA molecules present peptide antigens, the β2-microglobulin-associated CD1 molecules (CD1a, CD1b, CD1c, CD1d), expressed on professional antigen presenting cells (for example, dendritic cells), present lipids and glycolipids to T cells. CD1d-restricted, glycolipid-specific T-cell responses may be involved in the pathogenesis of autoimmune disease (Fig. 4). Most potentially pertinent to PNH is the finding that murine CD1d contains GPI in its presentation groove, and parasite GPI may be the target of protective T-cell immunity in murine malaria [42,43]. Thus, subject to direct proof that pluripotential HSCs express CD1d, CD1d-restricted GPI-specific T cells might be central to the pathogenesis of PNH.

Conclusion

In order to prove that in PNH there is conditional selection of the type we have outlined, the crucial test has two components. On one hand, one must demonstrate the presence in PNH patients of auto reactive T cells that recognize a target that is sharply differentially expressed on PNH HSCs *versus* non-PNH HSCs; on the other hand, one needs to identify the target itself, which is most likely a GPI-linked molecule or GPI itself. In general, the identification of antigen-specific T cells in clinical samples has relied on stimulating their growth by having the appropriate antigenic peptide presented by appropriate cells, with the read-out consisting in cytotoxicity, proliferation assays or cytokine secretion assays. A powerful advance has been the introduction of HLA class I-peptide tetrameric complexes. By using tetramer technology it is possible to identify, quantitate and sort CD8+ T cells specific for a certain antigen with a sensitivity as high as 0.01%. In our case, a problem with this approach is that, for the time being, we know of too many GPI-linked proteins expressed on mature blood cells, whereas we know very little about which ones are expressed on HSCs. On the other hand, GPI is a well defined unique candidate target, but MHC tetramers are not applicable to it. Very recently we have shown that CD1d tetramers can be constructed by the novel technique of oxidative refolding chromatography. We have validated the sensitivity and specificity of this technique by indentifying, *in vitro* as well as *ex vivo*, a CD1d-restricted subset of T cells (characterized by a NK T phenotype and a Vα24/Vβ11 TCR structure), specific for the glycolipid α-galactosylceramide (αGalCer) [24]. We expect it should be possible to construct CD1d tetramers with GPI (instead of αGalCer) in the presentation groove, and use them to stain lymphocytes from PNH patients and control individuals. If the role of these cells in the pathogenesis of PNH is then confirmed by functional assays, their selective removal might be of therapeutic value in PNH. Equally important, the same approach might be applicable to a certain proportion of patients with aplastic anemia.

Acknowledgements: This work was supported by the Associazione Italiana per la Ricerca sul Cancro (AIRC), the Aplastic Anemia Foundation of America New Researcher Award, the Norman and Rosita Winston and the Charles A Dana Foundations, and the National Institute of Health 5RO1 HL56778-03 Grant.

References

1. Hijmans van den Bergh AA. Ictère hémolytique avec crises hémoglobinuriques: Fragilité globulaire. *Rev. Méd.* 1911;31:63-69.
2. Marchiafava E, Nazari A. Nuovo contributo allo studio degli itteri cronici emolitici. *Policlinico (Sez. Med.)* 1911;18:241.
3. Crosby WH. Paroxysmal nocturnal hemoglobinuria: Plasma factors of the hemolytic system. *Blood* 1953;8:444-458.
4. Lewis SM, Dacie JV. The aplastic anaemia--paroxysmal nocturnal haemoglobinuria syndrome. *Br J Haematol* 1967;13:236-51.
5. Rotoli B, Luzzatto L. Paroxysmal nocturnal haemoglobinuria. *Baillieres Clin Haematol* 1989;2:113-38.
6. Rotoli B, Luzzatto L. Paroxysmal nocturnal hemoglobinuria. *Semin Hematol* 1989;26:201-7.
7. Takeda J, Miyata T, Kawagoe K, et al. Deficiency of the GPI anchor caused by a somatic mutation of the PIG-A gene in paroxysmal nocturnal hemoglobinuria. *Cell* 1993;73:703-11.
8. Bessler M, Mason PJ, Hillmen P, et al. Paroxysmal nocturnal haemoglobinuria (PNH) is caused by somatic mutations in the PIG-A gene. *Embo J* 1994;13:110-7.
9. Holguin MH, Fredrick LR, Bernshaw NJ, Wilcox LA, Parker CJ. Isolation and characterization of a membrane protein from normal human erythrocytes that inhibits reactive lysis of the erythrocytes of paroxysmal nocturnal hemoglobinuria. *J Clin Invest* 1989;84:7-17.
10. Young NS. The problem of clonality in aplastic anemia: Dr Dameshek's riddle, restated. *Blood* 1992;79:1385-92.
11. Luzzatto L, Bessler M, Rotoli B. Somatic mutations in paroxysmal nocturnal hemoglobinuria: a blessing in disguise? *Cell* 1997;88:1-4.
12. Nafa K, Bessler M, Castro-Malaspina H, Jhanwar S, Luzzatto L. The spectrum of somatic mutations in the PIG-A gene in paroxysmal nocturnal hemoglobinuria includes large deletions and small duplications. *Blood Cells Mol Dis* 1998;24:370-84.
13. Araten DJ, Nafa K, Pakdeesuwan K, Luzzatto L. Clonal populations of hematopoietic cells with paroxysmal nocturnal hemoglobinuria genotype and phenotype are present in normal individuals. *Proc Natl Acad Sci U S A* 1999;96:5209-14.
14. Rawstron AC, Rollinson SJ, Richards S, et al. The PNH phenotype cells that emerge in most patients after CAMPATH-1H therapy are present prior to treatment. *Br J Haematol* 1999;107:148-53.
15. Ware RE, Pickens CV, DeCastro CM, Howard TA. Circulating PIG-A mutant T lymphocytes in healthy adults and patients with bone marrow failure syndromes. *Exp Hematol* 2001;29:1403-9.

16.Araten DJ, Bessler M, McKenzie S, et al. Dynamics of hematopoiesis in paroxysmal nocturnal hemoglobinuira (PNH): no evidence for intrinsic growth advantage of PNH clones. *Leukemia* 2002 in press.
17.Luzzatto L, Bessler M. The dual pathogenesis of paroxysmal nocturnal hemoglobinuria. *Curr Opin Hematol* 1996;3:101-10.
18.Young NS, Maciejewski J. The pathophysiology of acquired aplastic anemia. *N Engl J Med* 1997;336:1365-72.
19.Karadimitris A, Notaro R, Koehne G, Roberts IA, Luzzatto L. PNH cells are as sensitive to T-cell-mediated lysis as their normal counterparts: implications for the pathogenesis of paroxysmal nocturnal haemoglobinuria. *Br J Haematol* 2000;111:1158-63.
20.Nagakura S, Ishihara S, Dunn DE, et al. Decreased susceptibility of leukemic cells with PIG-A mutation to natural killer cells in vitro. *Blood* 2002;100.
21.Pannetier C, Even J, Kourilsky P. T-cell repertoire diversity and clonal expansions in normal and clinical samples. *Immunol Today* 1995;16:176-81.
22.Karadimitris A, Manavalan JS, Thaler HT, et al. Abnormal T-cell repertoire is consistent with immune process underlying the pathogenesis of paroxysmal nocturnal hemoglobinuria. *Blood* 2000;96:2613-20.
23.Karadimitris A, Li K, Notaro R, et al. Association of clonal T-cell large granular lymphocyte disease and paroxysmal nocturnal hemoglobinuria (PNH): further evidence for a pathogenetic link between T cells, aplastic anaemia and PNH. *British Journal of Haematology* 2001;115:1-6.
24.Karadimitris A, Gadola S, Altamirano M, et al. Human CD1d-glycolipid tetramers generated by in vitro oxidative refolding chromatography. *Proc Natl Acad Sci U S A* 2001;98:3294-8.
25.Hiroshige Y, Matsumoto N, Harima K, Miwa S, Kamei T, Ishihara T. An autopsy case of aplastic anemia-PNH syndrome terminated in acute granulocytic leukemia. *Nippon Ketsueki Gakkai Zasshi* 1977;40:16-23.
26.Luzzatto L, Familusi JB, Williams CK, Junaid TA, Rotoli B, Alfinito F. The PNH abnormality in myeloproliferative disorders: association of PNH and acute erythremic myelosis in two children. *Haematologica* 1979;64:13-30.
27.Devine DV, Gluck WL, Rosse WF, Weinberg JB. Acute myeloblastic leukemia in paroxysmal nocturnal hemoglobinuria. Evidence of evolution from the abnormal paroxysmal nocturnal hemoglobinuria clone. *J Clin Invest* 1987;79:314-7.
28.Longo L, Bessler M, Beris P, Swirsky D, Luzzatto L. Myelodysplasia in a patient with pre-existing paroxysmal nocturnal haemoglobinuria: a clonal disease originating from within a clonal disease. *Br J Haematol* 1994;87:401-3.
29.Harris JW, Koscick R, Lazarus HM, Eshleman JR, Medof ME. Leukemia arising out of paroxysmal nocturnal hemoglobinuria. *Leuk Lymphoma* 1999;32:401-26.
30.Hillmen P, Lewis SM, Bessler M, Luzzatto L, Dacie JV. Natural history of paroxysmal nocturnal hemoglobinuria. *N Engl J Med* 1995;333:1253-8.
31.Brodsky RA, Vala MS, Barber JP, Medof ME, Jones RJ. Resistance to apoptosis caused by PIG-A gene mutations in paroxysmal nocturnal hemoglobinuria. *Proc Natl Acad Sci U S A* 1997;94:8756-60.
32.Ware RE, Nishimura J, Moody MA, Smith C, Rosse WF, Howard TA. The PIG-A mutation and absence of glycosylphosphatidylinositol-linked proteins do not confer resistance to apoptosis in paroxysmal nocturnal hemoglobinuria. *Blood* 1998;92:2541-50.

33.Chen R, Nagarajan S, Prince GM, et al. Impaired growth and elevated fas receptor expression in PIGA(+) stem cells in primary paroxysmal nocturnal hemoglobinuria. *J Clin Invest* 2000;106:689-96.

34.Kawagoe K, Kitamura D, Okabe M, et al. Glycosylphosphatidylinositol-anchor-deficient mice: implications for clonal dominance of mutant cells in paroxysmal nocturnal hemoglobinuria. *Blood* 1996;87:3600-6.

35.Rosti V, Tremml G, Soares V, Pandolfi PP, Luzzatto L, Bessler M. Murine embryonic stem cells without pig-a gene activity are competent for hematopoiesis with the PNH phenotype but not for clonal expansion. *J Clin Invest* 1997;100:1028-36.

36.Tremml G, Dominguez C, Rosti V, et al. Increased sensitivity to complement and a decreased red blood cell life span in mice mosaic for a nonfunctional Piga gene. *Blood* 1999;94:2945-54.

37.Bessler M, Mason P, Hillmen P, Luzzatto L. Somatic mutations and cellular selection in paroxysmal nocturnal haemoglobinuria. *Lancet* 1994;343:951-3.

38.Yamada N, Miyata T, Maeda K, Kitani T, Takeda J, Kinoshita T. Somatic mutations of the PIG-A gene found in Japanese patients with paroxysmal nocturnal hemoglobinuria. *Blood* 1995;85:885-92.

39.Nafa K, Bessler M, Deeg HJ, Luzzatto L. New somatic mutation in the PIG-A gene emerges at relapse of paroxysmal nocturnal hemoglobinuria. *Blood* 1998;92:3422-7.

40.Karadimitris A, Li K, Notaro R, et al. Association of clonal T-cell large granular lymphocyte disease and paroxysmal nocturnal haemoglobinuria (PNH): further evidence for a pathogenetic link between T cells, aplastic anaemia and PNH. *Br J Haematol* 2001;115:1010-4.

41.Karadimitris A, Luzzatto L. The cellular pathogenesis of paroxysmal nocturnal haemoglobinuria. *Leukemia* 2001;15:1148-52.

42. Naik RS, Branch OH, Woods AS, et al. Glycosylphosphatidylinositol anchors of Plasmodium falciparum: molecular characterization and naturally elicited antibody response that may provide immunity to malaria pathogenesis. *J Exp Med.* 2000;192:1563-1576.

43. Schofield L, Hewitt MC, Evans K, Siomos MA, Seeberger PH. Synthetic GPI as a candidate anti-toxic vaccine in a model of malaria. Nature. 2002;418:785-789.

A Possible Intrinsic Mechanism for Clonal Expansion of PNH Abnormal Cells

Norimitsu Inoue[1,2], Tomohisa Izui[2], Maki Kuwayama[2,3], Jun-ichi Nishimura[2,3], Ken Kurokawa[4], Takashi Machii[3], Yuzuru Kanakura[3] and Taroh .Kinoshita[2]

[1]Department of Molecular Genetics, Osaka Medical center for Cancer and Cardiovascular Diseases, Osaka, Osaka 537-8511, Japan
[2]Department of Immunoregulation, Research Institute for Microbial Diseases, Osaka University, Suita, Osaka 565-0871, Japan
[3]Department of Hematology and Oncology, Osaka University Medical School, Suita, Osaka 565-0871, Japan
[4]Genome Information Research Center, Osaka University, Suita, Osaka 565-0871, Japan

Summary: Paroxysmal nocturnal hemoglobinuria (PNH) is an acquired hematopoietic stem cell disorder that causes clonal expansion of Glycosylphosphatidyl-inositol (GPI)-anchor deficient cells. A somatic mutation of PIG-A that encodes a subunit of the enzyme complex in the GPI-anchor synthesis results in deficiency of GPI-anchored proteins. Two lines of evidence suggest that second abnormality is involved in clonal expansion of GPI-anchor deficient cells. Pig-a disrupted hematopoietic stem cells in mouse and PIG-A deficient cells in normal human individuals did not clonally expand. Two possible mechanisms for clonal expansion have been proposed. One is an immunoselection mechanism in that PNH cells are more resistant to immunological attack than normal cells because of the deficiency of GPI-anchored proteins. The other is an intrinsic mechanism in that PNH cells have a second mutation responsible for clonal expansion like benign tumors.

We reported that all of the GPI-anchor deficient cells in a patient with PNH carried both abnormality of chromosome 12, 46XX, t(12; 12)(q13; q15) and a mutation of PIG-A. We hypothesized that the abnormality of chromosome 12 is causally related to clonal expansion of PNH cells in this patient. In order to map the breakpoints and identify genes located there, we established hybrid cell lines by fusing GPI-anchor deficient monocytes of the patient with HPRT-deficient mouse myeloma cells and selected hybrid cell lines carrying each abnormal chromosome 12 by analyzing microsatellite markers. We determined the breakpoints by testing the presence of STS markers, Southern blotting and the inverse PCR method. We found that a 9.3Mbp fragment derived from 12q13-q15 region in one chromosome 12 was inserted into q15 region in the other chromosome 12. So, the chromosome abnormality was not a translocation but an insertion of a region from q13 to q15 and should be shown 46XX, ins(12; 12)(q15; q13q15).

We have identified a gene located in one of the breakpoints. I will discuss its possible involvement in clonal expansion of PNH abnormal cells.

Key words: PNH, Clonal expansion, Chromosomal abnormality, Intrinsic advantage, Hematopoietic stem cell

Introduction

PNH is an acquired hematopoietic stem cell disorder that causes clonal expansion of GPI-anchor deficient cells. A somatic mutation of PIG-A results in deficiency of GPI-anchor on hematopoietic stem cells in all patients with PNH[1].

PIG-A encodes an essential subunit of enzyme complex involved in a transfer of N-acetylglucosamine (GlcNAc) from UDP-GlcNAc to phosphatidylinositol at the first step of the GPI-anchor biosynthesis[2, 3]. The next important issue in pathogenesis of PNH is whether a mutation of PIG-A also causes clonal expansion of the GPI-anchor deficient hematopietic stem cells. Dr. Bessler's group and we have established Pig-a-disrupted mice by using Cre/loxP system and analyzed whether Pig-a disrupted hematopoietic stem cells clonally expand[4, 5]. These analyses showed that Pig-a disrupted hematopoietic stem cells did not expand relative to the normal cells. Furthermore Dr. Luzzatto's group found small population of PIG-A deficient hematopoietic cells in some normal individuals but these cells did not expand[6]. These results indicate that second abnormalities would mediate clonal expansion of GPI-anchor deficient cells. Immunological selection of GPI-anchor deficient cell has been proposed as one of the mechanisms for the clonal expansion. Bone marrow failure like aplastic anemia is a common feature in patients with PNH and thought to be induced by immunological attacks against hematopoietic stem cells. GPI-anchor deficient cells should be more resistant to the immunological attacks than normal cells because many GPI-anchored proteins are deficient. In fact, GPI-anchor deficient cells are detected in many patients with aplastic anemia[7, 8]. However, population of GPI-anchor deficient cells detected in aplastic anemia is smaller than that in de novo PNH and they do not increase quickly. In many cases with de novo PNH, most of the peripheral blood cells are occupied with GPI-anchor deficient cells. So, it is possible that PNH cells would clonally expand by one or more additional mutations introduced in GPI-anchor deficient cells themselves like benign tumors.

In this study, we show the first report of a second genetic abnormality, which occurred in GPI-anchor deficient cell of a patient with PNH.

Materials and Methods

FACS analysis

Cells were stained with anti-CD33 antibody, anti-CD9 antibody, anti-DAF antibody or anti-CD59 antibody directly conjugated with phycoerythrine or FITC (Pharmingen) in Phosphate buffered saline with 1% bovine serum albumin and 0.1% sodium azide. The stained cells were analyzed in a FACSCalibur (Becton Dickinson).

FACS sorting

Bone marrow cells were obtained from patient J20 with informed consent. Cells were stained with anti-CD59 antibodies. CD59-positive and –negative cell fractions were sorted with a FACStar (Becton-Dickinson) and the karyotype of each fraction was analyzed.

Detection of PIG-A mutations

PCR products of PIG-A gene amplified with primer sets as described previously[9] were analyzed with MDE (mutation detection enhancement) gel electrophoresis according to a protocol provided by a manufacturer. Abnormal PCR fragments were cloned into a plasmid vector and sequenced.

Establishment of hybrid cell lines

Mononuclear cell fraction was obtained from 25 ml of patient J20 blood sample by using Ficoll-Paque and reacted with anti-CD11b antibody conjugated magnetic beads (Miltenyi Biotec) for 15minutes on ice. Monocytes were selected from them in MACS system (Miltenyi Biotec). More than ninety percentages of these selected cells also expressed CD33. We fused 5×10^6 of these cells to 10^6 of hypoxanthine phosphoribosyl transferase (HPRT) negative mouse myeloma cell P3X63Ag8.563 with Hybri-Max (Sigma-Aldrich) and divided these fused cells into two 96-well plates. We cultured these hybrid cells in S-clone medium (Sanko Junyaku) with HAT (hypoxanthine, aminopterine and thymidine). When we expanded the cloned hybrid cells, we cultured them in RPMI1640 with 10% fetal bovine serum.

PCR

Ten thousand hybrid cells were suspended in 100μl of TE buffer and treated at 95 degree for 5 minutes. We used 2μl of these heat-treated cell samples in PCR reactions.

Analyses by microsatellite markers (D12S77 and D12S78) were performed with Texas red- conjugated and non-conjugated primers sets and the polymorphisms of PCR products were detected with Hitachi SQ-5500 sequencer.

Results and Discussions

GPI-anchor deficiency and chromosomal abnormality in patient J20

Patient J20 was diagnosed as aplastic anemia at 34 years of age and treated with anabolic steroids[10]. The treatment was effective for the pancytopenia. However, we detected microscopically a chromosomal translocation 46XX, t(12; 12)(q13; q15) in 50% of bone marrow cells (10/20) (Fig. 1).

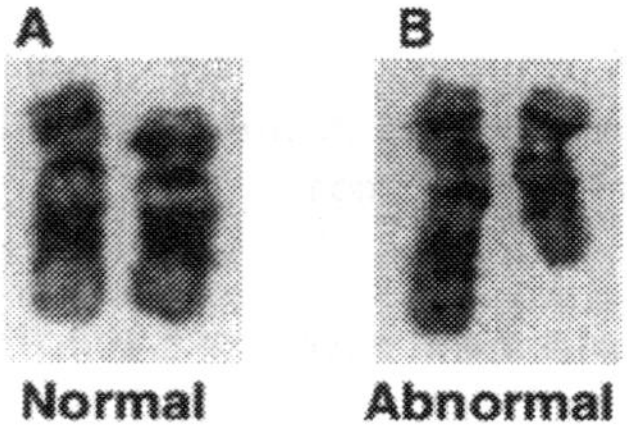

Fig. 1. Abnormality of chromosome 12 in patient J20. Her chromosomal abnormality was microscopically detected as 46XX, t(12;12)(q13;q15) in the bone marrow cells. Normal and abnormal chromosomes 12 are shown in A and B, respectively.

We also found deficiencies of GPI-anchored proteins, DAF and CD59 on her blood cells. The population of DAF- and CD59-deficient cells occupied 59.7% and 55.7% of erythrocytes, respectively and 23.4% and 24.2% of granulocytes, respectively. We, therefore, diagnosed PNH. First episode of hemoglobinuria was observed six months after GPI-anchor deficient cells were detected. Both the bone marrow cells carrying the chromosomal translocation (16/20) and the GPI-anchor deficient peripheral blood cells (Fig.2) were increasing gradually. The percentages of CD59-deficient cells were 75%, 78% and 100% in erythrocytes, polymorpho nuclear cells (PMN) and monocytes, respectively. However we did not detect the deficiency of the surface expression of the GPI-anchored proteins on B and T lymphocytes unlike other patients with PNH.

The percentage of the GPI-anchor deficient erythrocytes or granulocytes is similar to that of cells carrying the translocation. To examine whether all of GPI-anchor deficient cells have this chromosomal abnormality, we sorted CD59-positive and -negative bone marrow cells by a FACStar (Becton-Dickinson) and analyzed karyotypes of each fraction of sorted cells. Karyotypes of all CD59 positive bone marrow cells were normal (20/20), while those of CD59 negative cells were abnormal (18/18), indicating that only GPI-anchor deficient cells should have this chromosomal abnormality. We predicted that the chromosomal abnormality would be involved in deficiency of the GPI-anchor synthesis instead of a mutation of PIG-A or that it would be the second mutation in addition to a mutation of PIG-A and induce clonal expansion of the GPI-anchor deficient cells.

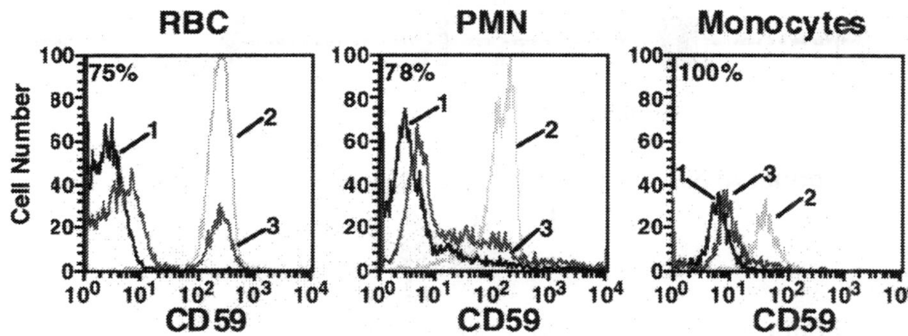

Fig. 2. The expression of CD59 on peripheral blood cells of patient J20. Erythrocyte (left panel), polymorpho nuclear cells (PMN) (middle panel) and monocytes (right panel) of a normal individual (lines 2) and patient J20 (lines 3) are stained with phycoerythrine-conjugated anti-CD59 antibody. The antibody was not added in negative controls of cells of patient J20 (lines 1).

A mutation of PIG-A

To identify whether mutation of PIG-A gene causes the deficiency of GPI-anchor in this patient, we performed heteroduplex analysis with MDE gel electrophoresis by using genomic DNA from her granulocytes. We detected abnormal mobility in a region including a junction of exon 2 and intron 2 (data not shown). This abnormality was derived from a nucleotide substitution from Guanine to Adenine at codon 239 that would generate amino acid exchange from Glycine to Arginine (Fig 3). Two mutations in codon 239 have been reported to cause a loss of the function of PIG-A[11]. As this mutation was at the 3' end of exon 2, it is possible that it might affect the splicing to generate a premature stop codon or induce a decrease of amount of the transcript. This result suggests that a mutation of PIG-A but not the chromosomal abnormality is involved in the deficiency of the GPI-anchor similarly to other patients and that the chromosomal abnormality should be related to clonal expansion of the GPI-anchor deficient cells in this patient because two rare independent mutations occurred in only one bone marrow cell. We started to identify the breakpoints in the abnormal chromosomes in order to clone genes responsible for clonal expansion of the GPI-anchor deficient cell.

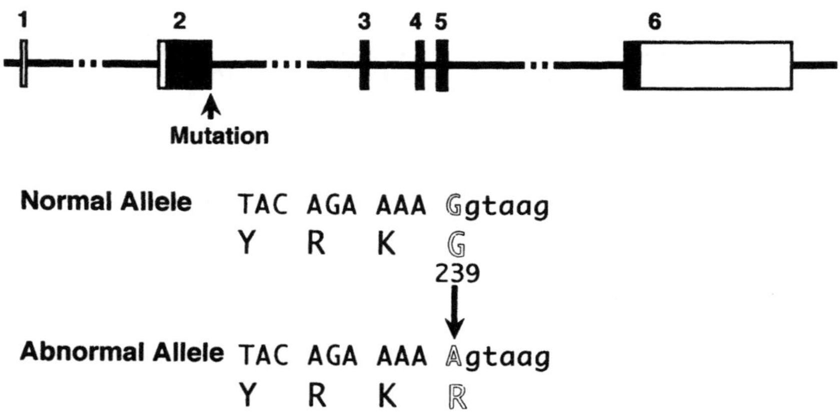

Fig. 3. A somatic mutation of *PIG-A* gene in patient J20. This patient had a nucleotide exchange from Guanine to Adenine at the 3' end of exon 2 of *PIG-A* gene. This mutation would generate an amino acid exchange from Glycine to Arginine at codon 239.

Establishment of hybrid cell lines

To analyze the chromosomal abnormality, it is necessary to establish cell lines derived from GPI-anchor deficient cell fraction of this patient. Because she does not have any GPI-deficient lymphoid cells, we fused her monocytes to HPRT negative mouse myeloma cell line to establish hybrid cells containing abnormal chromosomes. We collected monocytes from her mononuclear cell fraction using anti-CD11b antibody conjugated MACS magnetic beads, fused them to myeloma cell line and selected hybrid cells by medium with HAT. Cells containing chromosome 12 from patient J20 were chosen based on the expression of CD9 whose gene is mapped to chromosome 12. To characterize the breakpoints, we selected hybrid cell lines containing either one of the abnormal chromosomes 12 by PCR using two microsatellite markers, D12S77 (Fig. 4) and D12S78 (data not shown). Using genomic DNA of these clones, we determined the breakpoints. We found that a 9.3Mbp fragment derived from 12q13-q15 region in one chromosome 12 was inserted into q15 region in the other chromosome 12. So, the chromosome abnormality was not a translocation but an insertion of a region from q13 to q15 and should be shown 46XX, ins(12; 12)(q15; q13q15).

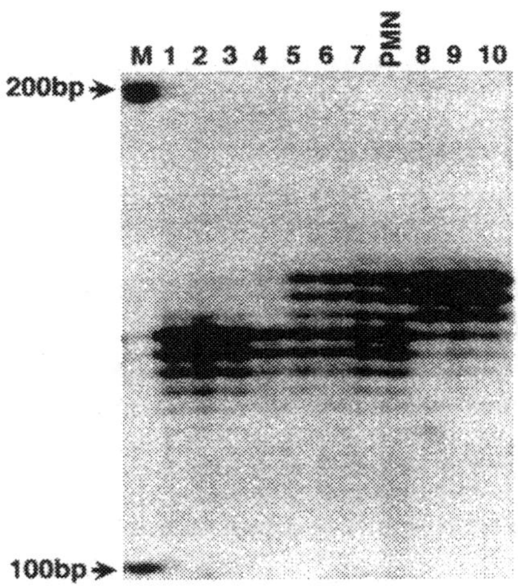

Fig. 4. Characterization of hybrid cell lines. Hybrid cell lines were established by fusion of monocytes from patient J20 with mouse myeloma cell line. PCR to detect microsatellite marker D12S77 was performed by using DNA of hybrid clones. It was indicated that hybrid clones 1-4 and 8-10 have different chromosomes 12 and hybrid clones 5-7 have both chromosomes 12. PMN was used as a positive control. M; molecular marker

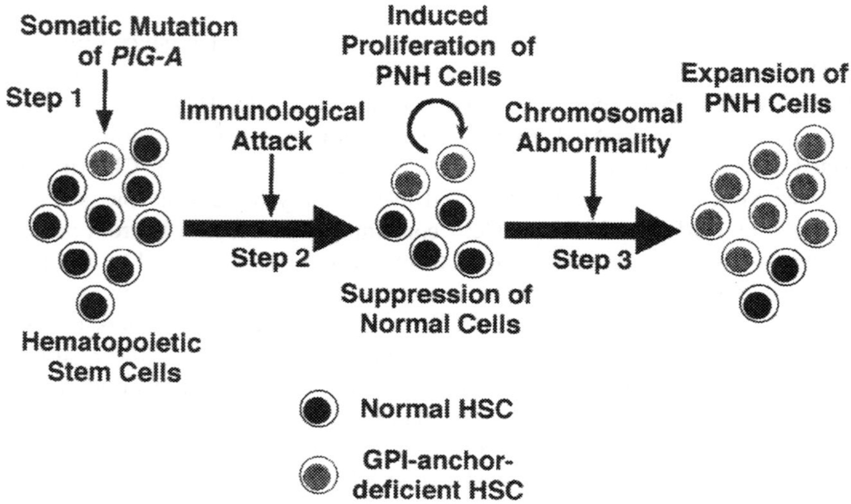

Fig. 5. Predicted mechanism for clonal expansion of PNH cells in patient J20. HSC; Hematopoietic stem cells

Predicted mechanism for clonal expansion

In this study, we showed that the second rare mutation (the abnormality of chromosome 12) occurred in the hematopoietic stem cells containing PIG-A mutation. The population of GPI-anchor deficient cell was gradually increasing with the appearance of the chromosomal abnormality. This result suggests that one or more of the genes mapped to these breakpoints should be involved in the clonal expansion of the GPI-anchor deficient cells.

We show that predicted molecular mechanism for clonal expansion of PNH cells in patient J20 (Fig. 5). In the first step, a somatic mutation of PIG-A occurs in one of the hematopoietic stem cells. By unknown immunological mechanism in aplastic anemia, proliferation and development of normal hematopietic stem cells are suppressed (step 2). On the other hand, GPI-anchor deficient cells derived from cells containing a mutation of PIG-A should escape from the immunological attack and proliferate to support the hematopoiesis. At this time, we can detect the GPI-anchor deficient cells but these cells don't occupy all of the hematopoiesis over normal cells.

The proliferation of these cells increases their opportunities for receiving other mutations. Then the second abnormality occurs and induces clonal expansion of these cells like benign tumors (step 3). If small mutation can induce a similar effect to the chromosomal abnormality, the similar molecular mechanism would account for clonal expansion in many other patients with PNH. We should analyze the second mutation in other patients with PNH based on the abnormality in patient J20.

Acknowledgments

We thank Fumiko Ishii, Keiko Kinoshita, Reiko Fukuyama and Keiko Yamamoto for the excellent technical assistance. This study is supported by grants from the Ministry of Education, Culture, Sports, Science and Technology of Japan.

References

1. Kinoshita T, Inoue N, Takeda J (1995) Defective glycosyl phosphatidylinositol anchor synthesis and paroxysmal nocturnal hemoglobinuria Adv. Immunol. 60:57-103
2. Watanabe R, Kinoshita T, Masaki R, et al. (1996) PIG-A and PIG-H, which participate in glycosylphosphatidylinositol anchor biosynthesis, form a protein complex in the endoplasmic reticulum. J. Biol. Chem. 271:26868-26875
3. Watanabe R, Inoue N, Westfall B, et al. (1998) The first step of glycosylphosphatidylinositol biosynthesis is mediated by a complex of PIG-A, PIG-H, PIG-C and GPI1. EMBO J 17:877-885
4. Tremml G, Dominguez C, Rosti V, et al. (1999) Increased sensitivity to complement and a decreased red cell life span in mice mosaic for a non-functional Piga gene. Blood 94:2945-2954
5. Murakami Y, Kinoshita T, Maeda Y, et al. (1999) Different roles of glycosylphosphatidylinositol in various hematopoietic cells as revealed by model mice of paroxysmal nocturnal hemoglobinuria. Blood 94:2963-2970
6. Araten D J, Nafa K, Pakdeesuwan K, et al. (1999) Clonal populations of hematopoietic cells with paroxysmal nocturnal hemoglobinuria genotype and phenotype are present in normal individuals. Proc Natl Acad Sci USA 96:5209-5214
7. Dunn D E, Tanawattanacharoen P, Boccuni P, et al. (1999) Paroxysmal nocturnal hemoglobinuria cells in patients with bone marrow failure syndromes. Ann Intern Med 131:401-408

8. Wang H, Chuhjo T, Yamazaki H, et al. (2001) Relative increase of granulocytes with a paroxysmal nocturnal haemoglobinuria phenotype in aplastic anaemia patients: the high prevalence at diagnosis. Eur. J. Haematol. 66:200-205
9. Iida Y, Takeda J, Miyata T, et al. (1994) Characterization of genomic PIG-A gene:a gene for GPI-anchor biosynthesis and paroxysmal nocturnal hemoglobinuria Blood 83:3126-3131
10. Nishimura J-I, Inoue N, Azenishi Y, et al. (1996) Analysis of PIG-A gene in a patient who developed reciprocal translocation of chromosome 12 and paroxysmal nocturnal hemoglobinuria during follow-up of aplastic anemia. Am J Hematol 51:229-233
11. Luzzatto L, Nafa K (2000) Genetics of PNH. In: N. S. Young and J. Moss (eds) Paroxysmal nocturnal hemoglobinuria and the glycosylphosphatidylinositol-linked proteins. Academic Press, San Diego, pp21-47

Part 5

PNH Clones Under Bone Marrow Failure

Clinical Significance of Increased PNH-Type Cells in the Peripheral Blood of Patients with Aplastic Anemia and Refractory Anemia

Shinji Nakao, Hong-bo Wang, Tatsuya Chuhjo
Cellular Transplantation Biology, Kanazawa University Graduate School of Medical

Science, 13-1 Takaramachi, Kanazawa, Ishikawa 920-8641, Japan

Summary: A small number of blood cells with the paroxysmal nocturnal hemoglobinuria (PNH) phenotype are occasionally detected in patients with aplastic anemia (AA) and those with myelodysplastic syndrome (MDS). In an attempt to determine the clinical significance of PNH cells in bone marrow failure syndrome, we examined peripheral blood red blood cells (RBCs) and granulocytes of AA patients and MDS/refractory anemia (RA) patients, for the presence of $CD55^{-}CD59^{-}$ (PNH) cells using a sensitive flow cytometry. A total of 100 AA patients and 105 RA patients with varying disease duration were studied. A significant increase (≥0.003%) in the percentage of PNH cells compared to normal individuals was detected in 52% of AA patients. The frequency of patients with increased PNH cells was particularly high as 88.6% in newly diagnosed AA patients. On the other hand, the increased PNH cells were detected only in 15 of 105 (14.3%) RA patients. None of the 15 patients with increased PNH cells showed apparent trilineage dysplasia; in these patients, RA was diagnosed mainly based on the presence of erythroid dysplasia and/or mild trilineage dysplasia as well as hypercellularity in the bone marrow aspirates from the sternum or the iliac bone. None of them exhibited karyotypic abnormalities. Pancytopenia of these patients improved with cyclosporine or anabolic steroids, or remained stable without treatments. Thus, the presence of PNH cells in peripheral blood probably serves as a marker for good prognosis in RA patients.

Altogether, these findings indicate that detecting a small number of PNH-type cells using flow cytometry is crucial in assessing pathophysiology of bone marrow failure syndrome such as AA and RA as well as choosing an appropriate therapy.

Key words: aplastic anemia, paroxysmal nocturnal hemoglobinuria, refractory anemia, flow cytometry

Introduction

Paroxysmal nocturnal hemoglobinuria (PNH) is an acquired hemolytic anemia characterized by an increase in the number of glycosylphosphatidylinositol (GPI)-anchored protein-deficient cells in periheral blood. The GPI-anchored protein-negative cells are derived from hematopoietic progenitor cells with a mutation of PIG-A gene. Recent studies demonstrated that a small number of such PNH-type cells are present in peripheral blood of normal individuals due to somatic mutation of PIG-A gene in hematopoietic progenitor cells [1]. PNH is thought to be a result of predominant clonal proliferation of hematopoietic progenitor cells with PIG-A mutation although the mechanisms of clonal expansion are unknown.

An increase (>1%) in the proportion of PNH-type cells has been demonstrated in patients with AA [2,3] and those with myelodysplastic syndrome (MDS) [4,5]. The prevalence of increased PNH cells reportedly varies from 20% to 50%. A study from NIH demonstrated that an increase (>1%) of PNH granulocytes defined as $CD16^-CD66^-CD15^+$ cells were detected in 22% of patients with AA before treatment as well as in 23% of patients with myelodysplastic syndrome (MDS), and that MDS patients with increased PNH granulocytes were likely to respond to immunosuppressive therapy compared to those not showing increased PNH granulocytes, though the study did not find this correlation with response to immunosuppressive therapy in AA patients [6]. In recent years, PNH cells less than 1% of total granulocytes or red blood cells (RBCs) can be reliably detected using two color flow cytometry [1]. Such a minor population of PNH cells in bone marrow failure patients may have a clinical significance different from that of PNH cells exceeding 1% of total granulocytes or RBCs.

We then examined peripheral blood RBCs and granulocytes of AA patients and MDS/refractory anemia (RA) patients, for the presence of $CD55^-CD59^-$ (PNH) cells using a sensitive flow cytometry.

Material and methods

Patients

Three to 20 ml of heparinized blood was drawn from 68 normal individuals including 42 males and 26 females, aged 22-71 years old with a median age of 48, 100 patients with AA including 35 untreated patients who were newly diagnosed, and 105 patients with RA. Ham's test and the sugar-water test were negative in all patients. All patients provided informed consent prior to sampling, and the study was approved by the institutional human research committee.

Antibodies and flow cytometry

We used phycoerythrin (PE)-labeled anti-CD11b monoclonal antibodies (MoAb, Becton Dickinson, Mountain View, California, USA), PE-labeled anti-glycophorin A MoAb, isotype-matched control MoAb (Becton Dickinson), fluorescein-isothiocyanate (FITC)-labeled anti-CD55 (clone IA10, mouse IgG2a, Pharmingen, San Diego, CA, USA), FITC-labeled anti-CD59 (clone p282, mouse

IgG2a, Pharmingen) and isotype-matched control MoAb (Pharmingen). For granulocyte analysis, blood was subjected to RBC lysis using ammonium chrolide buffer and leukocytes were incubated with PE-labeled anti-CD11b MoAb and FITC-labeled anti-CD55 and anti-CD59 MoAb or isotype matched control MoAb on ice for 25 minutes. For RBC analysis, blood was washed two times with PBS and incubated with PE-labeled anti-glycophorin A MoAb and FITC-labeled anti-CD55 and anti-CD59 MoAb. A total of at least 10^5 cells within the gate corresponding to granulocyte or RBCs were analyzed by FACScan (Becton Dickinson, Mountain View, California, USA). The percentage of $CD55^-CD59^-CD11b^+$ cells among total $CD11b^+$ cells and that of $CD55^-CD59^-$ glycophorin A^+ cells among total glycophorin A^+ cells were determined for each individual.

Immunosuppressive therapy

Antithymocyte globulin (ATG, Lymphoglobuline, Merieux, Lyon, France, 15 mg/kg/d, 5 days) and cyclosporine A (CyA, Novartis, 6 mg/kg/d, 6 months) were administered to 26 patients with AA after analysis of PNH granulocytes. Response was evaluated at 6 months after therapy. Response criteria included resolution of requirement for transfusions and 2 g/dl or more rises of hemoglobin levels.

Results

Sensitivity of the flow cytometry for detecting PNH granulocytes in normal individuals

To test validity of the flow cytometric assay reported by Araten *et al* (13), 10^6 granulocytes of 5 normal individuals were examined for the presence of PNH granulocytes. All 5 individuals showed 0.001 to 0.002% of $CD55^-CD59^-CD11b^+$ within the granulocyte gate. Most $CD55^-CD59^-$ cells in the CD11b negative fraction proved to be cell debris since they are located in the left lower position in the forward scatter/side scatter dot plot graph. This result is consistent with that reported by the above authors. When blood from a PNH patient was diluted with blood from a normal individual and the percentage of PNH cells in 10^5 $CD11b^+$ cells was determined, this flow cytometry could detect as few as 0.002% PNH granulocytes as estimated. Since most of the study subjects had profound granulocytopenia at sampling, 10^5 cells in the granulocyte gate were analyzed in the following experiments.

Comparison between detecting PNH granulocytes and PNH RBCs

Although detection of a minor PNH granulocyte population using the two color flow cytometry is a reliable method, minor PNH-cell populations smaller than 0.1% are often difficult to assess. Detection of RBCs defective of both CD55 and CD59 in addition to PNH granulocytes may substantiate a diagnosis of bone marrow failure with a minor PNH-cell population. We then studied peripheral blood of 10 AA patients with a minor population of PNH granulocytes for the presence of PNH RBCs. **Fig. 1** shows representative histograms. All patients exhibited a distinct population of CD55$^-$CD59$^-$ cells in RBCs which were much easier to discriminate from normal cells than in granulocytes. Therefore, in recent analysis of RA patients, we studied both granulocytes and RBCs to detect a minor population of PNH-type cells with a greater sensitivity.

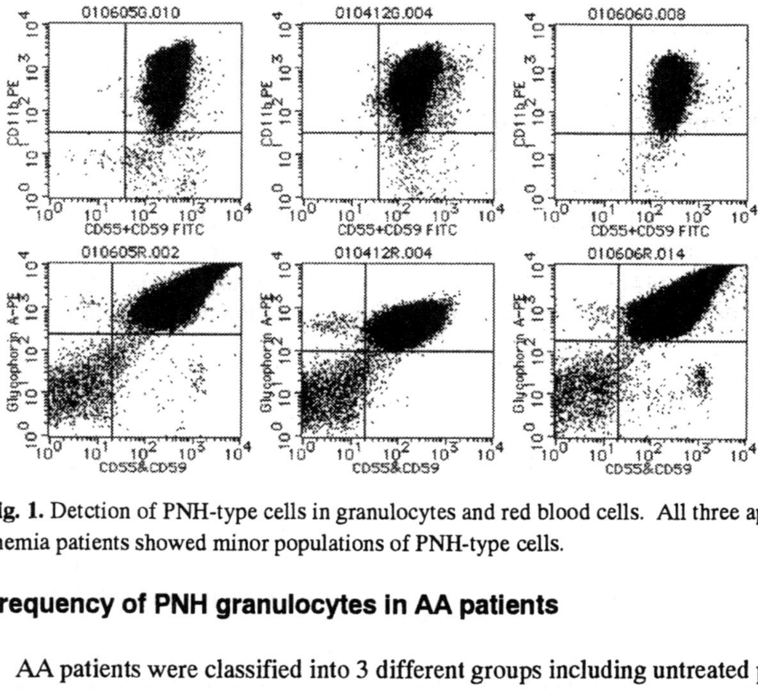

Fig. 1. Detction of PNH-type cells in granulocytes and red blood cells. All three aplastic anemia patients showed minor populations of PNH-type cells.

Frequency of PNH granulocytes in AA patients

AA patients were classified into 3 different groups including untreated patients at diagnosis (35), treated patients with a short (<5 years) disease duration (16) and those with a long (>5 years) disease duration (49). **Fig. 2** illustrates frequencies of PNH granulocytes among 10^5 CD11b$^+$ granulocytes in normal individuals and each patient group. A small number of CD55$^-$CD59$^-$CD11b$^+$ cells were detected in 18 of 68 normal individuals though its frequency did not exceed 0.003% in any of them. In untreated AA patients who were newly diagnosed, 31 (88.6%) patients exhibited greater than 0.003% of PNH granulocytes. Representative histograms showing varying frequencies of PNH granulocytes in 4 newly diagnosed patients are shown in **Fig. 3**. The proportion of patients showing an increase (>0.003%) of PNH granulocytes in treated patients with a short and long disease duration was 68.6% (11/16) and 20.4% (10/49), respectively. There was a significant difference in the proportion of patients showing increased PNH granulocytes between AA patients at diagnosis (p<0.0001) or treated patients with a short disease duration (p=0.001) and those with a long disease duration.

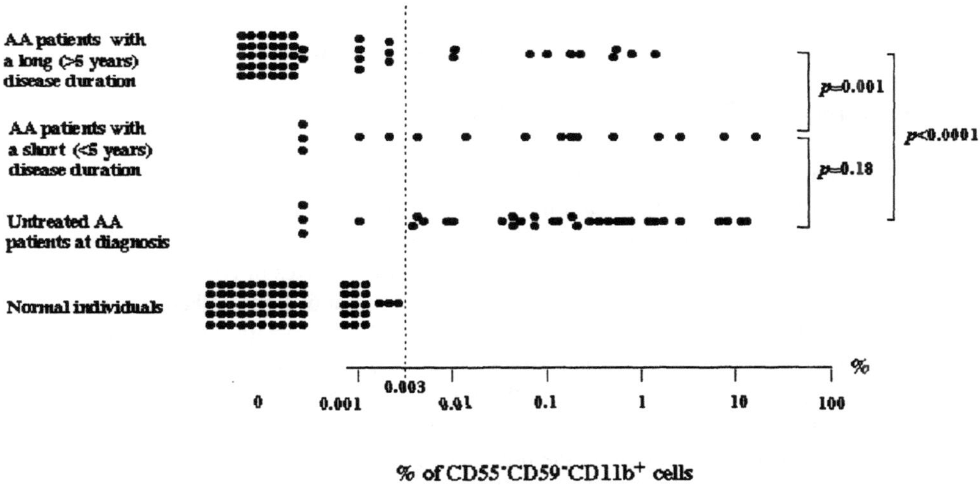

Fig. 2. Percentages of PNH granulocytes among 10^5 $CD11b^+$ granulocytes in normal individuals and aplastic anemia patients. The dotted line represents a value of 0.003%.

Fig.3.

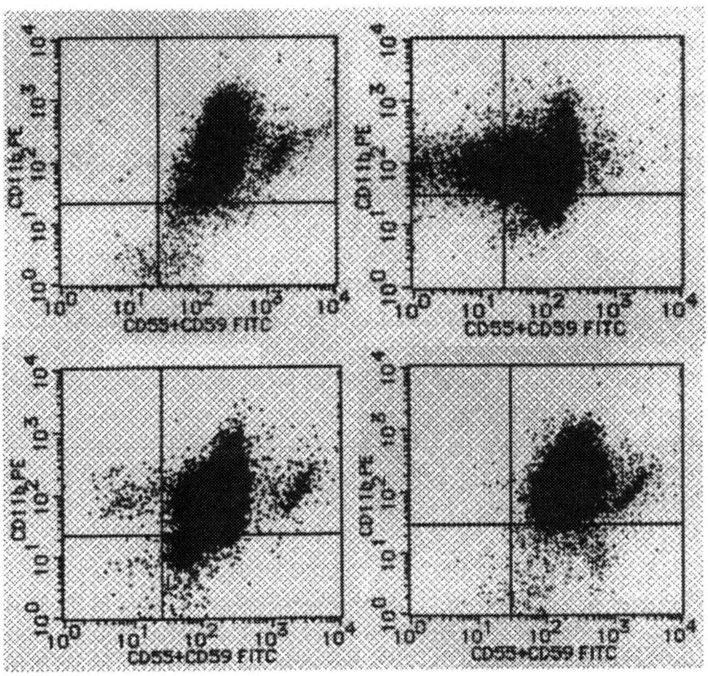

Fig. 3. Detection of PNH granulocytes in the peripheral blood of newly diagnosed alastic anemia patients. Histograms of 4 representative cases are shown.

Frequency of PNH cells in RA patients

In contrast to the high prevalence of patients with a minor population of PNH cells in AA patients, a significant increase of PNH cells was detected in only 14 (15%) of 105 RA patients studied. The percentage of PNH cells in the positive patients ranged from 0.01% to 1.5% of the total granulocytes or RBCs and it was less than 1.0% in 12 (80%) of them. This small number of PNH cells had been considered to be insignificant by previous studies. To see whether the RA patients with an increased PNH cells share some common features, we reviewed clinical records and bone marrow pictures of these patients. None of the patients with increased PNH cells showed apparent trilineage dysplasia; in these patients, RA was diagnosed mainly based on the presence of erythroid dysplasia and/or mild trilineage dysplasia as well as hypercellularity in the bone marrow aspirates from the sternum or the iliac bone. None of them exhibited karyotypic abnormalities. Clinical course of this subset of RA patients was distinct from the other RA patients; pancytopenia was not progressive or spontaneously resolved in 4 patients. Seven other patients responded to cyclosporine (3), anabolic steroids (2), or menatetrenone (2) and became free of transfusions. Response to cyclosporine in the remaining 2 patients was too early to assess.

The relative increase of PNH cells was not detected in any of 4 patients with RA with ringed sideroblasts, 30 patients with RA with excess of blast (RAEB), and 6 patients with RAEB in transformation.

Changes in the frequency of PNH granulocytes associated with immunosuppressive therapy

Peripheral blood of 19 untreated patients with increased PNH granulocytes was reexamined at 6-12 months after ATG plus CyA therapy. Fifteen of 18 (83.3%) patients who responded to the immunosuppressive therapy showed a decrease in the frequency of PNH granulocytes to 0.01–90% of pretreatment values. The frequency slightly increased in the remaining 3 recovering patients and in one patient refractory to the therapy.

Discussion

A sensitive flow cytometric analysis used in this study showed that a small number of granulocytes deficient of CD55 and CD59 are present in the peripheral blood of approximately 26.5% (18/68) of normal individuals. The frequency of PNH granulocytes did not exceed 0.003% of total granulocytes. These results were similar to those reported by Araten *et al* [1] in which PNH cells were detected at an average frequency of 0.0022% in 9 normal individuals. This frequency is of the same order of magnitude as the estimated mutation frequency in normal individuals in other genes such as Hprt and glycophorin A [7].

Although many studies have documented the presence of PNH cells in the peripheral blood of AA patients, the number of studies focusing on newly diagnosed AA patients is limited. Schrezenmeier *et al* [8] reported that an increase in the frequency of PNH cells was detected in 11 of 21 (52%) AA patients analyzed before first treatment and that those patients with increased PNH cells were less likely to respond to immunosuppressive therapy compared to AA patients without increased PNH cells. De Lord *et al* [9] also found increased frequency of PNH cells in 30% (6/20) of untreated AA patients although they did not observe any correlation between increased PNH cells and response to immunosuppressive therapy. More recently, Dunn *et al* [6] reported that 15% (5/33) of AA patients analyzed before any definitive therapy showed an increase of PNH granulocytes. All of these previous studies defined 1 or more percentages of PNH cells being abnormal based on their analysis of normal individuals. Our study showed an increase in the frequency of PNH granulocytes (>0.003%) in 88.6% of AA patients who were newly diagnosed and had not undergone definitive therapy. Previous reports appear to have underestimated the proportion of patients showing increased PNH granulocytes perhaps due to a lower sensitivity of the flow cytometric assay used.

In contrast to the high prevalence of patients with increased PNH cells in AA patients, such patients were detected only 14% of RA patients. The percentage was lower than that of a similar study using insensitive methods by Dunn et al (30%) [5] and was comparable to that reported by Iwanaga et al (13%) [5]. The difference between out study and others may be due to differences in clinical backgrounds including disease duration or in diagnostic criteria for MDS that involve subjective judgment. Our findings indicate that AA is a different disease from MDS in terms of possessing increased PNH cells.

Previous studies suggest that successful treatment of AA might result in expansion of PNH cells, based on high incidence of PNH in patients in remission after immunosuppressive therapy [10,11]. We thus expected that an increase in the frequency of PNH granulocytes would be detected more often in treated patients than in those newly diagnosed. Contrary to expectation, the proportion of patients with increased PNH granulocytes in patients treated with immunosuppressive therapy 5 or more years previously was lower than those of untreated patients and treated patients with a short disease duration. Moreover, the frequency of PNH granulocytes in newly diagnosed patients decreased in association with haematologic recovery after immunosuppressive therapy in most patients probably due to an increase of $CD55^{+}CD59^{+}$ granulocytes. A large body of evidence suggests that PNH stem cells may acquire survival advantages in the presence of immunologic pressure on haematopoietic stem cells due to a deficiency of GPI-anchored proteins required for being attacked by cytotoxic T cells [12,13]. Taken together, these suggest that cell-mediated attack against haematopoietic cells is probably involved in the pathogenesis of AA in most patients.

Detecting a minor population of PNH cells in RA patients has an important implication in the clinical practice because patients with increased PNH cells showed clinical features similar to moderate AA rather than to true MDS. Some of these patients had been treated by cytotoxic therapy such as low-dose cytosine arabinoside without effects before undergoing immunosuppressive therapy. Previous reports on successful therapy of MDS-RA with cyclosporine or ATG may have included patients with increased PNH cells that could be diagnosed as AA rather than MDS. The presence of PNH cells in peripheral blood probably serves as a marker for good prognosis in RA patients.

Altogether, these findings indicate that detecting a small number of PNH-type cells using flow cytometry is crucial in assessing the pathophysiology of bone marrow failure syndrome such as AA and RA as well as choosing an appropriate therapy.

References

1. Araten DJ, Nafa K, Pakdeesuwan K, et al (1999) Clonal populations of hematopoietic cells with paroxysmal nocturnal hemoglobinuria genotype and phenotype are present in normal individuals. Proc Natl Acad Sci U S A 96: 5209-5214
2. Schubert J, Vogt HG, Zielinska-Skowronek M, et al (1994) Development of the glycosylphosphatitylinositol-anchoring defect characteristic for paroxysmal nocturnal hemoglobinuria in patients with aplastic anemia. Blood 83: 2323-2328
3. Azenishi Y, Ueda E, Machii T, et al (1999) CD59-deficient blood cells and PIG-A gene abnormalities in Japanese patients with aplastic anaemia. Br J Haematol 104: 523-529
4. Nagakura S, Kawaguchi T, Fujimoto K, et al (1997) Sequential development of myelodysplasia and paroxysmal nocturnal hemoglobinuria in a patient with preceding aplastic anemia. Int J Hematol 65: 187-189
5. Iwanaga M, Furukawa K, Amenomori T, et al (1998) Paroxysmal nocturnal haemoglobinuria clones in patients with myelodysplastic syndromes. Br J Haematol 102: 465-474
6. Dunn DE, Tanawattanacharoen P, Boccuni P, et al (1999) Paroxysmal nocturnal hemoglobinuria cells in patients with bone marrow failure syndromes. Ann Intern Med 131: 401-408
7. Albertini RJ, Nicklas JA, O'Neill JP, et al (1990) In vivo somatic mutations in humans: measurement and analysis. Annu Rev Genet 24: 305-326
8. Schrezenmeier H, Hertenstein B, Wagner B, et al (1995) A pathogenetic link between aplastic anemia and paroxysmal nocturnal hemoglobinuria is suggested by a high frequency of aplastic anemia patients with a deficiency of phosphatidylinositol glycan anchored proteins. Exp Hematol 23: 81-87.
9. De Lord C, Tooze JA, Saso R, et al (1998) Deficiency of glycosylphosphatidyl inositol-anchored proteins in patients with aplastic anaemia does not affect response to immunosuppressive therapy. Br J Haematol 101: 90-93

10. Socie G, Henry-Amar M, Bacigalupo A, et al (1993) Malignant tumors occurring after treatment of aplastic anemia. European Bone Marrow Transplantation-Severe Aplastic Anaemia Working Party. N Engl J Med 329: 1152-1157
11. De Planque MM, Kluin-Nelemans HC, van Krieken HJ, et al (1988) Evolution of acquired severe aplastic anaemia to myelodysplasia and subsequent leukaemia in adults. Br J Haematol 70: 55-62.
12. Young NS (1992) The problem of clonality in aplastic anemia: Dr Dameshek's riddle, restated. Blood 79: 1385-1392
13. Takami A, Zeng W, Wang H, et al (1999) Cytotoxicity against lymphoblastoid cells mediated by a T-cell clone from an aplastic anaemia patient: role of CD59 on target cells. Br J Haematol 107: 791-796

Incidence and Clinical Significance of PNH clone in Myelodysplastic Syndromes

Masao Tomonaga, Masako Iwanaga, Kengo Fuchigami, Yoriko Inoue, Tatsuro Joh, Itsuro Jinnai

Department of Hematology, Molecular Medicine Unit, Atomic Bomb Disease Institute, Nagasaki University School of Medicine, Sakamoto 1-12-4, Nagasaki, Japan

Summary: We have detected PNH clone in four (10%) out of 40 cases of primary MDS by flowcytomery using monoclonal antibodies against CD59 and CD55. All of four cases had PIG-A mutation(s) that were heterogeneous and similar to those reported in de novo PNH and aplastic anemia/PNH. Two cases of MDS/PNH responded to immunosuppressive therapy (IST) with cyclosporin A; the PNH clone persisted even after obtaining a good response. Clonal analysis for one case of MDS/PNH with 8 trisomy by combining FACS cell-sorting and FISH revealed that the PNH clone had developed from MDS clone. Another clonal analysis in one case of AML M6 transformed from de novo PNH also revealed that M6 had originated from the PNH clone. Thus, our study indicates that (1) PNH clone can develop from MDS, (2) MDS/PNH case can respond to immunosuppressive therapy and (3) MDS/AML can originate from de novo PNH clone. However, recent literature suggests that MDS/AML clone develops not only from a PNH clone as in our cases, but also from a residual stem cell in a minority of PNH cases. Therefore, molecular genetic studies are essential to compare stem cells of PNH clone and those of non-PNH clone in respect of immune surveillance and leukemogenesis.

Key words: primary MDS, PNH clone in MDS, MDS/AML from de novo PNH

Introduction

Classical de novo PNH is molecularly defined by PIG-A mutation of blood cells and clinically by intravascular hemolysis and thrombosis. PNH hematopiesis is clonal, although the underlying process for its growth advantage over normal hematopoiesis is not fully understood [1-3]. PNH is also known to progress to aplastic anemia (PNH/AA), and rarely if not at all transforms to MDS/AML [4-8]. Conversely, idiopathic AA develops PNH clone in 20~40%(AA/PNH) as well as MDS clone in 5~15% during natural course of the disease or after immunosuppressive therapy (IST) [9, 10]. The PNH clones associated with AA also carry similar PIG-A mutations[11, 12] and AA/PNH also not infrequently transforms to MDS/AML [13-17]. In contrast, PNH clone associated with MDS has been less studied for its incidence, PIG-A mutation and clinical significance [18, 19]. A rare case of PNH associated with myelofibrosis that transfomed to AML was also reported [20].

The rate of transformation to MDS/AML from de novo PNH has been reported extremely low, ~5% [1, 21, 22]. Some of these MDS/AML cases have clonal origin from the PNH clone [4-7], but al least one case of de novo PNH [8] and two cases of AA/PNH [15-17] seemed to have MDS or MDS/AML clone derived from a non-PNH clone. Such a residual stem cell may be preleukemic in nature.

Recently we have elucidated the incidence of PNH clone in a series of 40 patients with primary MDS and analysed molecularly CD59-deficient granulocytes for the presence of PIG-A mutations [18]. We have also conducted immunosuppressive therapy with cyclosporin A (CSA) in 15 patients with primary MDS including two cases with a PNH clone. In addition we performed a clonal analysis in one case of AML M6 transformed from de novo PNH. Based on our observations and recent literature, we discuss on the clinical and biological significance of PNH clone in MDS and in general.

Methods

Detection of PNH clone by flowcytometry [18]

For analysis of GPI-anchor protein-deficient red blood cells and granulocytes anti-CD59 and anti-CD55 monoclonal antibodies were used. In the early phase of our study to elucidate incidence of PNH clone in a series of 40 patients with MDS, we performed flowcytometry as single-labeling with either of the above antibodies for cells gated on the basis of side scatter, and in the late phase we employed double or triple labeling, such as CD11b, CD59 and CD55 for granulocytes. We considered that sensitivity of our flowcytometry was rather crude as more than a few % in the former, and more than 0.1% in the latter.

Detection of PIG-A mutations in PNH clones [18]

Genomic DNA was extracted from peripheral blood granulocytes. PCRs were performed with six specific primer sets. The amplified PCR products were then applied for single-strand conformation polymorphism (SSCP) screening. The PCR products were directly inserted into a TA cloning vector. The recombinant plasmids were prepared from transformed E. Coli.. Cycle sequencing reactions were performed using Taq Dye Deoxy Terminator Cycle Sequencing kit.

Sequencing was performed using an automated sequencer (ABI model 377; Perkin-Elmer).

To identify abnormal splicing caused by somatic mutations of PIG-A coding region, we analysed RT-PCR-Southern blot with RNA of granulocytes. RT-PCR was performed and the identity of the product was confirmed by Southern blot hybridization with full-length PIG-A cDNA that was kindly provided by Dr. J. Takeda (Osaka University) and analyzed by autoradiography. The largest RT-PCR product in each sample was cloned into the vector pBluescript II^R and sequenced as described above.

Combined FACS and FISH analysis in MDS/PNH with 8 trisomy [18]

In a case (Pt. Y.N.) of MDS with a PNH clone and 8 trisomy in 2 (10%) of 20 bone marrow metaphases, FISH for interphase analysis was performed using a chromosome 8 probe on FACS-sorted peripheral blood granulocytes;$CD16^{dim}$ and $CD16^{bright}$ populations. For hybridization we used biotin-labeled DNA probes specifying for the alpha-satellite of chromsome 8 (D8Z2, Oncor) and then incubated with FITC-labeled avidin (Oncor).

Follow-up study of PNH clone during immunosuppressive therapy

15 patients of MDS including two cases with a PNH clone were treated with CSA $5mg/M^2$/day for more than 6 months. Clinical response was graded into good response, partial response, minimal response and no response according to the criteria by Japan Intractable Disease Study Group.

Clonal analysis of blast cells in AML transformed from de novo PNH

A 56 year-old male case of de novo PNH was diagnosed to have MDS after 17 years of stable PNH state with frequent hemolysis. Within a few months blast cell outgrowth became evident and a diagnosis of AML M6 (erythroleukemia) was established. He was treated with an intensive chemotherapy consisting of daunorubicin and cytarabine. Complete remission was obtained, but morphological dyplasia as well as CD59-deficient granulocytes and red blood cells were persisting. The CD13-positive blast cells of M6 phase were tested for the presence of CD59 and CD55-deficiency by double-staining flowcytometry.

Results

Incidence of PNH clone in MDS

Among 40 patients with primary MDS, PNH cells negative for CD59 or CD55 were detected in four patients (Table 1). In these patients with MDS/PNH, the populations of granulocytes and erythrocytes negative for CD59 were smaller than those of de novo PNH. Remaining MDS patients did not show any CD59-negative population at a sensitivity of a few % or more. There was no correlation between the duration from the initial diagnosis of MDS and the incidence of PNH clone. All four MDS/PNH patients were free from transfusion and treatment with CSA at the time of PNH clone analysis. One of the MDS/PNH patient had 8 trisomy in 2 (10%) of 20 bone marrow metaphases.

Table 1. CD59-negative cells in patients with MDS/PNH and de novo PNH

Patients	Type	% affected RBC	Ham's test	%affected granullocytes
MDS/PNH				
Pt. I.S.	III	12.2	+	15.6
Pt. K.T.	III	17.5	+	95
Pt. N.T.	III	21.2	+	19.5
Pt. N.Y.	III	21.5	+	18.3
De novo PNH				
Pt. U.N.	II+III	80.0	+++	100
Pt. U.T.	III	80.0	+++	80.0
Pt. K.K.	II+III	34.9	+++	53.4

PIG-A mutation in MDS/PNH clones

The results of PIG-A mutation analysis were summarized in table 2. PIG-A mutations were found in granulocytes of all patients with CD59-negative granulocytes irrespective of MDS/PNH or de novo PNH. In MDS/PNH, three of four cases had two independent clones of different PIG-A mutations, whereas one of three de novo cases had two clones. In type and site, PIG-A mutations were heterogeneous, indicating absence of mutation specific to MDS. These findings were consistent with the previous reports of patients with de novo PNH and AA/PNH.

Table 2. Mutations of PIG-A gene in patients with MDS/PNH and de novo PNH

Patient	Type	Position	Consequence	Mutant/ total clones
MDS/PNH				
Pt. I.S.	Insertion	exon 3 751T	Frameshift(299)	3/15
	Deletion	exon 3 729T	Framesgift(245)	2/15
Pt. K.T.	Deletion	exon 2 700C	Frameshift(179)	4/18
	Base change	exon2 214C to T	His(72)-T	6/15
Pt. N.T.	Base change 3'	splice site of intron4 ag/G to gg/G	Deletion exon5	2/9
	Base change	exon 6 1321T to C	Phe(441)-Leu	2/9
Pt. Y.N.	Base change	exon 4 908A-G	His(303)-Arg	2/11
De novo PNH				
Pt. U.N.	Deletion	exon5 1097T	Frameshift(379)	8/14
Pt. U.T.	Deletion exon5 &5'	splice site of intron5 1187~88AG/g	Deleltion exon5	8/12
Pt. K.K.	Deletion exon3	776T	Frameshift(259)	8/10

8 trisomy clone and PNH clone share a common population in a case of MDS/PNH

We have set the cut-off limit at 0.7% based on our preliminary tests for normal granulocytes from healthy controls. By cell sorting CD16^{bright} granulocytes (non-PNH clone) and CD16^{dim} (PNH clone) were isolated. These populations were analysed by FISH for two signals (disomy) and three signals (8 trisomy). 8 trisomy cells were found in 13/500 (3%) of the CD-16^{dim} population, but only in 2/500 (0.4%) in CD16^{bright} population as shown in Table 3.

PNH clones persist after responding to immunosuppressive therapy

Flowcytometric analysis for CD55/59-negative granulocytes was performed in 9 patients with MDS who responded to CSA. In two cases with MDS/PNH (Pt.Y.N. in Table 1 and a newly diagnosed patient T.K.). In Pt. Y.N. the PNH clone was decreased in percentage but apparently persisting after responding to CSA. In Pt. T.K. PNH clone was not measured but Ham's test was negative before CSA, but 12% positive when Hb elevated during IST and increased to 54 % after IST began to lose its effect. In the remaining 7 cases without PNH clone before starting IST no newly appearing PNH clone was detected after responding to CSA by the most sensitive flowcytometry.

Table 3. Interphase FISH for CD16^{dim}-PNH population in patients with MDS/PNH with 8 trisomy

Sorted cells	No. cells with 3dotts	total	% cells with 3 dotts
CD16^{dim} population	13	500	2.6
CD16 ^{bright} population	2	500	0.4

Cut-off point for normal granulocytes in three healthy controls:0.7%

Table 4. Follow-up study for PNH clone in patients with MDS and MDS/PNH that responded to immunosuppressive therapy (IST) with cyclospsoroin A

Patients	% CD-59 negative cells (Hb;g/dl)			
	Before IST	During IST	Follow-up	Intervals(year)
MDS/PNH				
Pt. Y.N.	18 (6.0)	-	8.9 (10.8)	5
Pt. T.K.	Ham (-) (6.5)	12 (8.7)	54 (7.5)	6.5
MDS 7 cases	0 (6.0 median)		0 (11.6 median)	3.5 (median)

AML M6 blasts derived from de novo PNH clone

The AML M6 case transformed from de novo PNH had 5% myeloblasts that were demonstrated CD13- and CD34-positive by flowcytometry. Erythroblasts comprised 80%. These CD13-positive myeloblasts were 96.5%CD55-negative and 88.8%CD59-negative, indicating that AML M6 clone had derived from de novo PNH clone in this case.

Table 5. Clonal analysis for CD13-positive myeloblasts in a case of AML tansformed from de novo PNH

Cells	CD55		CD59	
	%Negative	%Positive	%Negative	%Positive
CD13-positive blasts	96.5	3.5	88.8	11.2

Discussion

In the present survey for 40 patients with primary MDS including 31 refractory anemia(RA) cases, four patients (10%) had definite PNH clone(2). All of them were RA ; the incidence being 13% in RA. Regarding the incidence of PNH clone in primary MDS, Dunn et al. reported a much higher incidence of 23%[19]. More recently, using a highly sensitive flowcytomeric analysis Nakao et al [this proceeding] reported that ~80% of AA patients and ~14% of MDS patients had a PNH clone. The incidence of PNH clone in MDS may be lower than that in AA. The PNH clone in MDS seems to be less proliferative than those in de novo PNH and some cases of AA/PNH.

Our study indicated for the first time that PIG-A mutations also occur in primary MDS. There was neither hot spot nor specific mutations in MDS, similar to de novo PNH and AA/PNH. Three out four MDS/PNH cases had two independent clones of distict mutations, whereas we observed such independent clones in one out of three de novo PNH cases. These findings indicate that the PIG-A gene is highly mutable in these bone marrow failure syndromes.

It is important to elucidate whether MDS clone itself is the origin of a secondary PNH clone(s) or a residual stem cell out side MDS clone gives rise to a PNH clone(s). In this context, our combined FACS/FISH analysis in a case of MDS/PNH with 8 trisomy proved that PNH clone originated from MDS clone. Table 6 summarizes 12 reports in the literature which studied clonal relationship between PNH and MDS or AML. In our case of AML M6 transformed from de novo PNH the CD13(+) blasts in AML M6 phase were derived fromne PNH clone.

Table 6. Clonal studies on MDS/AML in patients with de nvo PNH, AA/PNH and MDS

Reported by	Disease	Evolution	Clonal relation
Devine [4]	de novo PNH	AML	same(CD55-)
Parlier [5]	de novo PNH	RARS	probably same
Shichishima [6]	de novo PNH	AML	same (CD55-)
Nakahata [20]	PNH with fibrosis	AML	same (CD55-)
Longo [7]	de novo PNH	RARS	same (all CD59-)
Nishimura [13]	AA/PNH	Translocation 1	same (CD59-)
Nagakura [14]	AA	MDS & PNH	probably same
Iwanaga [18]	MDS	PNH	same (FACS/FISH)
Tomonaga [this study]	de novo PNH	AML	same (CD55-)
Van Kamp [8]	de novo PNH	MDS	separate (CD16+)
Jin [15]	AA/PNH	MDS	separate (CD55+)
Mortazavi [16]*	AA/PNH	MDS/AML	separate (all CD55+)
Ishihara [17]	AA/ PNH(2 cases)	MDS/AML	separate (all CD55+)

*same case as in Jin

There have been three similar reports of same clonal relation [4, 6, 20]. However, interestingly there were two reports that indicated an appearance of AML in a separate clone outside PNH clone [16, 17]. These two cases were AA/PNH and developed AML through MDS phase. Also there are two reports of same clonal origin of MDS [5, 7] and one report of separate clonal origin of MDS in de novo PNH [8]; thus MDS may also develop either from original PNH clone or from a separate clone (stem cell) outside PNH clone. We should suspect that not only the PNH clone but also the residual stem cells may have predisposition toward leukemogenesis. It should be emphasized that there have been a scarcity of clonal study concerning clonal origin of MDS or AML in de novo PNH, since such MDS/AML transformation is a rare event, probably less than 5% of ceses of de novo PNH.

Regarding pathology of the residual stem cells in de novo PNH, there have been two recent reports indicating presence of qualitative abnormality in cellular function [23, 24]. Chen et al. reported that CD34(+)CD59(-) PNH stem cells had near normal growth capacity in the presence of multiple cytokines in liquid culture, but CD34(+)CD59(+) residual stem cells had a profound depression in growth capacity as well as increased susceptibility to FAS receptor-mediated apoptosis[23]. Maciejewski et al. of NIH group also observed that not only CD34(+)CD59(-) PNH stem cells but also CD34(+)CD59(+) residual stem cells showed impaired cloning efficiency as measured by in vitro colony formation and LT-CIC assay [24]. In this respect Young et al. pointed out by comparing the above two reports and other reports that there are still a lot of discrepancy in terms of growth capacity and apoptosis of PNH stem cells and residual stem cells, suggesting difference in assay systems employed [25].

Recently IST is one of the choices for the therapy of MDS, especially for RA. At our institute 15 cases of MDS, mostly RA including two cases of MDS/PNH have been treated with CSA. 9 cases (60%) showed good response in hemoglobin value with an increment of more than 2 g/dl. Response in platelet count and neutrophil count was apparently inferior to Hb. This is a shrap contrast to AA which usually shows balanced response in trilinegaes. Interestingly one of two patients with MDS/PNH and 8 trisomy responded to CSA very well and 8 trisomy. disappeared during IST, but the PNH clone was persisting.

Thus we confirmed that IST is also effective in MDS including MDS/PNH cases. Dunn et al. also reported much higher response rate for MDS cases with PNH clone compared with cases without [19]. Whether de novo PNH responds to IST is another important question. Some cases of de novo PNH, especially those with hypoplastic nature did respond to IST, but more cases of typical de novo PNH with cellular marrow and overt hemolysis did not [26, 27]. This is a sharp contrast to the generally high response rate (~80%) of AA. Regarding immunosuppression to hematopoietic stem cells in these bone marrow failure syndromes, we need to conduct a larger scale comparative clinical trial using the same IST protocol to reveal similarity or dissimilarity in response rate and degree between de novo PNH and PNH/AA group, AA and AA/PNH group, and MDS and MDS/PNH group by using highly sensitive flowcytometry for the detetction of PNH clone.

In summary, secondary PNH clone can originate from primary MDS clone and does not seem to expand markedly but persists even after obtaining hematological response to IST. MDS and AML may originate not only from PNH clone but also from a residual stem cell outside PNH clone. IST seems a valuable tool to elucidate a further interesting pathology related to immune suppression of stem

cells by immunocompetent cells; high response rate to IST in AA, and moderate and incomplete response in MDS. In this respect we need to accumulate more clinical data about the efficacy of IST for de novo PNH.

References

1. Hillmen P, Lewis SM, Bessler M et al (1995) Natural history of paroxysmal nocturnal hemoglobinuria. Br J Haematol 333:1253- 1258
2. Rosse WF, Ware RE (1995) The molecular basis of paroxysmal nocturnal hemoglobinuria. Blood 86:3277-3286
3. Hillmen P, Richards SJ (2000) Implications of recent insights into pathophysiology of paroxysmal nocturnal hemoglobinuria. Br J Hematol 108:470-479
4. Devine DV, Gluck WL, Rosse WF et al (1987) Acute myeloblastic leukemia in paroxysmal nocturnal hemoglobinuria. Evidence of evolution from the abnormal paroxysmal nocturnal hemoglobinuria clone. J Clin Invest 79:314-317
5. Parlier V, Tiainen M, Beris et al (1992) Trisomy 8 in granulocytic, erythrocytic and megakaryocytic lineages by chromosomal in situ suppression hybridization in a case of refractory anemia with ringed sideroblasts comlicating the course of paroxysmal nocturnal hemoglobinuria 81:296-304
6. Shichishima T, Terasawa T, Hashimoto C et al (1993) Discordant and heterogeneous expression of GPI-anchored membrane proteins on leukemia cells in a patient with paroxysmal nocturnal hemoglobinuria. Blood 81:1855-1862
7. Longo L, Bessler M, Beris P et al (1994) Myelodysplasia in a patient with pre-existing paroxysmal nocturnal hemoglobinuria: a clonal disease originating within a clonal disease. Br J Haematol 87:401-403
8. Van Kamp H, Smit JW, van den Berg E et al (1994) Myelodysplasia following paroxysmal nocturnal hemoglobinuria: evidence for the emergence of separate clone. Br J Haematol 87:399-400
9. Socie G, Rosenfeld S, Frickhofen N et al (2000) Late clonal diseases of treated aplastic anemia. Sem in Hematol 37:91-101
10. Tooze JA, Marsh JC, Gordon-Smith EC (1999) Clonal evolution of aplastic anemia to meylodysplasia/acute myeloid leukemia and paroxysmal nocturnal hemoglobinuria. Leuk Lyphoma 33:231-241
11. Nagarajan S, Brodsky RA, Young NS et al (1995) Genetic defects underlying paroxysmal nocturnal hemoglobinuria that arises out of aplastic anemia. Blood 86:4656-4661
12. Azenishi Y, Ueda E, Machii T et al (1999) CD59-deficient blood cells and PIG-A gene abnormalities in Japanese patients with aplastic anemia. Br J Haematol 104:523-529
13. Nishimura JI, Inoue N, Azenishi Y et al (1996) Analysis of PIG-A gene in a patient who developed reciprocal translocation of chromosome 12 and paroxysmal nocyurnal hemoglobinuria during follow-up of aplastic anemia. Am J Hematol 51:229-233
14. Nagakura S, Kawaguchi T, Fujimoto K et al (1997) Sequential development of myelodysplasia and paroxysmal nocturnal hemoglobinuria in a patient with preceding aplastic anemia. Int J Hematol 65:187-189

15. Jin JY, Tooze JA, Marsh et al (1996) Myelodysplasia following aplastic anemia-paroxysmal nocturnal hemoglobinuria syndrome after treatment with immunosuppression and G-CSF:evidence for the emergence of a separate clone. Br J Haematol 94:510-512
16. Mortazavi Y, Tooze JA, Gordon-Smith EC et al (2000) N-RAS gene mutation in patients with aplastic anemia and aplastic anemia/paroxysmal nocturnal hemoglobinuria during evolution to clonal disease. Br J Haematol 95:646-650
17. Ishihara S, Nakakuma H, Kawaguchi T (2000) Two cases showing clonal progression with full evolution from aplastic anemia–paroxysmal nocturnal hemoglobinuria syndrome to myelodysplasia and leukemia. Int J Hematol 72: 206-209
18. Iwanaga M, Furukawa K, Amenomori T et al (1998) Paroxysma nocturnal hemoglobinuria clones in patients with myelodysplastic syndromes. Br J Haematol 102:465-474
19. Dunn DE, Tanawattanacharoen P, Boccuni P et al (1999) Paroxysmal nocturnal hemoglobinuria cells in patients with bon marrow failure symdromes. Annal Int Med 131:401-408
20. Nakahata J, Takahashi M, Fuse I et al (1993) Paroxysmal nocturnal hemoglobinuria with myelofibrosis: progression to acute myelobalstic leukemia. Leuk Lymphoma 12:137-142
21. Harris JW, Koscick R, Lazaarus HM et al (1999) Leukemia arising out of paroxysmal nocturnal hemoglobinuria. Leuk Lymphoma 32: 401-406
22. Cornelis F, Motfort L, Osselaer JC et al (1996) Acute leukaemia in paroxysmal nocturnal hemoglobinuria. Case report and review of the literture. Hematol Cell Ther 38:285-288
23. Chen R, Nagarajan S, Orince GM et al (2000) Impaired growth and elevated fas receptor expression in PIGA(+) stem cells in primary paroxysmal nocturnal hemoglobinuria. J Clin Invest 106:689-696
24. Maciejewski J, Sloand E, Sato T et al (1997) Impaired hematopoiesis in paroxysmal nocturnal hemoglobinuria is not associated with a selective proliferative defect in the glycoslphosphatidylinositol-anchored protein-deficient clone. Blood 89:1173-1181
25. Young NS, Maciejewski JP (2000) Genetic and environmental effects in paroxysmal nocturnal hemoglobinuria: this little PIG-A goes " why?why?why?. J Clin Invest 106:689-696
26. Paquette RL, Yoshimura R, Veiseh C et al (1997) Clinica characteristics predict response to antithymocyte globulin in paroxysmal nocturnal hemoglobinuria. Br J Haematol 96:92-97
27. Stoppa AM, Vey N, Sainty D et al (1996) Correction of aplastic anemia complicating paroxysmal nocturnal hemoglobinuria; absence of eradication of the PNH clone and dependence of response on cyclosporin A administration. Br J Haematol 93:42-4
28. Van Kamp H, van Imhoff GW, de Wolf JT et al (1995) The effect of cyclosporine on haematological parameters in patients with paroxysmal nocturnal hemoglobinuria. Br J Haematol 89:79-82
29. Schubert J, Vogt HG, Zielinska-Skowronek M et al (1994) Development of the glycosylphosphatidyl-inositol-anchoring defect characteristic for paroxysmal nocturnal hemoglobinuria in patients with aplastic anemia. Blood 83:2323-2328

THE ENIGMA OF THE APLASTIC ANEMIA/PNH SYNDROME

Neal S. Young, Jaroslaw P. Maciejewski, Elaine Sloand, Guiben Chen, Weihua Zeng, Antonio Risitano, and Akira Miyazato

Hematology Branch, National Heart, Lung, and Blood Institute, Building 10, Room 7C103, NIH, Bethesda MD 20892-1652.

Summary: Bone marrow failure has been regarded as one consequence of paroxysmal nocturnal hemoglobinuria (PNH), and PNH in turn has been described as a late event in patients recovering from aplastic anemia. Better understanding of the pathophysiology of both diseases and improved tests for cell surface glycosylphosphatidylinositol (GPI)-linked proteins have radically altered this view. Flow cytometry of granulocytes shows evidence of an expanded PNH clone in a large proportion of marrow failure patients at the time of presentation: in our large NIH series, about 1/3 of over 200 aplastic anemia cases and almost 20% of more than 100 myelodysplasia cases. Clonal PNH expansion (rather than bone marrow failure) is strongly linked to the histocompatability antigen H.A.-DR2 in all clinical varieties of the disease, suggesting a relationship to the immune pathophysiology. An extrinsic mechanism of clonal expansion is also more consistent with knock-out mouse models and culture experiments with primary cells and cell lines, which have failed to demonstrate an intrinsic proliferative advantage for PNH cells. DNA chip analysis of multiple paired normal and PIG-A mutant cell lines and lymphoblastoid cells do not show any consistent differences in levels of gene expression. In aplastic anemia/PNH there is surprisingly limited utilization of the V β chain of the T cell receptor, and patients' dominant T cell clones, which are functionally inhibitory of autologous hematopoiesis, use identical CDR3 regions for antigen binding. Phenotypically normal cells from PNH patients proliferate more poorly in culture than do the same patient's PNH cells, and the normal cells are damaged as a result of apoptosis and overexpress Fas. Differences in protein degradation might play a dual role in pathophysiology, as GPI-linked proteins lacking an anchor would be predicted to be processed by the proteasome machinery and displayed in a class I HLA. context, in contrast to the normal pathway of cell surface membrane recycling, lysosomal degradation, and presentation by class II HLA. In summary, the strong relationship of a chronic, organ-specific immune destructive process and the expansion of a single mutant stem cell clone remains frustratingly enigmatic but likely to be the result of unusual and interesting biology.

Key words. hematopoiesis, aplastic anemia, bone marrow failure, autoimmunity, T cells

Introduction

Paroxysmal nocturnal hemoglobinuria (PNH) came to medical attention because of its unusual clinical presentation of pigment loss in the urine (the long history of the investigation of the disease has been eloquently recounted by Rosse (1)). "Intermittent hematinuria" was later recognized as due to the destruction of red cells in the circulation, and the unusual intravascular hemolysis shown to be secondary to the susceptibility of red cells to a non-antibody component of perturbed serum preparations. When in vitro hemolysis on exposure to acidified serum was utilized to formulate the diagnostic Ham test, the symptoms and signs of the disease were broadened and PNH was more fully characterized by a clinical triad of hemolysis, venous thrombosis, and marrow failure. Indeed, thrombotic disease, which is often in unusual anatomical locations, recurrent, and resistant to therapy, is the major cause of death among Western patients. In contrast, pancytopenia predominates as a manifestation of PNH in Asian countries. PNH is closely related to the disease aplastic anemia, in which immune-mediated hematopoietic cell destruction results in empty bone marrow (2): historically, patients with aplastic anemia have appeared to evolve to PNH, with evidence of hemolysis combined with a positive Ham test appearing months or years after successful treatment with immunosuppresive drugs (2).

The full characterization of the somatic genetic defect in PNH and its biochemical consequences were required to elucidate the extraordinarily close relationship between PNH and aplastic anemia. The mechanism of red cell lysis by treated sera was shown to be activation of the alternative pathway of complement (3). The erythrocyte defect was identified functionally as an inability to inactivate complement on the cell surface, in turn due to a deficit in specific membrane proteins (decay accelerating factor [DAF or CD55] and membrane inhibitor of reactive lysis [MIRL or CD59]). It was especially puzzling that not only were these proteins absent but PNH cells lacked a wide variety of others, including leucocyte alkaline phosphatase and erythrocyte acetylcholinesterase. Ultimately, demonstration of the enzymatic release of alkaline phosphatase by a specific phospholipase led to the identification of the common structural feature of a highly diverse group of missing proteins--a distinctive biochemical linkage to the cell membrane, the glycosylphosphatidylinositol (GPI) anchor. Preformed GPI is attached covalently to proteins that bear the appropriate carboxyl sequence and links them to the plasma membrane through the phosphatidylinositol moiety. GPI-anchored proteins also appear to cluster on the cell surface in biophysically distinctive regions, insoluble in certain detergents, termed rafts. Almost 100 different GPI-anchored proteins have been identified to date in mammalian cells, including enzymes, adhesion molecules, receptors, and blood group antigens (4) The GPI anchor is evolutionarily conserved, and most parasite proteins use this linkage (5). The biosynthesis of the GPI anchor was determined for trypanosomes and later in deficient mammalian cell lines (6). The genetic defect in PNH, discovered by expression cloning and then phenotypic correction of mutant cell lines, was localized to the *PIG-A* (phosphatidylinositol glycan-class A) gene (7); *PIG-A* encodes a protein required for the transfer of N-acetylglucosamine to phosphatidylinositol, an early step in the synthesis of the anchor (3). Immortalized cell lines from PNH patients all belong to the class A complementation group; the most frequent *PIG-A* alterations are small deletions that create stop codons or frameshifts, followed by missense mutations and small insertions.

PNH is a hematologic disease, and to date *PIG-A* gene mutations have only been described in hematopoietic cells. The cellular clonal origin of PNH was first inferred from G6PD enzyme analysis of circulating red cells in informative heterozygous females, and by implication all defective erythrocytes, leucocytes, and platelets originated from a single hematopoietic progenitor (8). Indeed, in PNH patients GPI-anchored proteins are missing from variable proportions of erythrocytes, granulocytes, platelets, and some lymphocytes, consistent with the activity of an aberrant hematopoietic stem cell clone. Because of the X chromosome location of *PIG-A*, a mutation is dominant at the cellular level due to Lyonization in women and hemizygosity in men.

PNH/Aplastic Anemia

Analysis of GPI-anchored proteins by flow cytometry is both more accurate and sensitive than the Ham test, and analysis of granulocytes avoids the obvious problem related to the presence of normal donor erythrocytes in patients dependent on red blood cell transfusions. From its first application in aplastic anemia (9), flow cytometry has revealed an expanded PNH clone in a very high proportion of patients; in the most recent analysis of our large series of over 200 cases, PNH was present in 42% (10). Even this high number may be an underestimate, as the distribution of GPI-anchored protein-negative cells in the "normal" range was skewed rightwards for PNH patients. Additionally, expansion of GPI-anchored protein-negative cells is almost always apparent at the time of diagnosis of aplastic anemia, and thus PNH is only unusually a "late" clonal event, as was incorrectly inferred from Ham test results. Indeed, about the same number of patients acquired a clone as showed spontaneous disappearance of aberrant cells. Furthermore, PNH clonal expansion is also associated with the myelodysplastic syndromes, and was present in 18% of 120 of our NIH patients (mainly referred at early FAB stages for the diagnosis and treatment of pancytopenia, rather than progression to leukemia) (10).

Flow cytometric testing indicates that marrow failure and PNH clonal expansion frequently coexist. Functional studies also show similarity between these syndromes. Aplastic anemia is characterized by a markedly diminished proliferative capacity of the marrow, as assayed in a variety of in vitro culture systems, including assays in semisolid medium for myeloid (CFU-GM), erythroid (BFU-E, CFU-E), megakyaryocytic (CFU_{Meg}), and multipotential (CFU_{GEMM}) progenitor cells; primitive progenitors (LTC-IC); and CD34 cell growth in long term bone marrow cultures. Although the morphology of the marrow in PNH is often relatively cellular, especially demonstrating erythroid hyperplasia, results in colony assays are similar to those observed in overt aplastic anemia (11-13). Thus patients with PNH have evidence of a functional defect in blood cell production similar to other marrow failure syndromes.

Pathophysiology of PNH Clonal Expansion

Can the genetics of PIG-A explain the pathophysiology of PNH? Is acquisition of the somatic mutation sufficient to explain all the manifestations of the disease, especially the marrow failure component? In the simplest model, PIG-A deficiency could confer on mutant cells an intrinsic growth or survival advantage that would lead inevitably to clonal expansion. Clinical observations and most of the experimental data fail to support this hypothesis. Most obviously, PNH as a disease does not behave like a leukemia, with inexorable proliferation of abnormal cells; in most patients, the proportion of PNH cells is stable over years and their differentiation proceeds along the usual lineage pathways. PNH is a rare disease, but normal individuals harbor tiny PNH clones: very small numbers, 20-60/million, of GPI-anchored protein-deficient, PIG-A- granulocytes were detected after repeated cell sorting of blood of healthy volunteers (14). In chimeric knock-out mice, PIG-A- cells also constitute only a minor, static fraction of the marrow or circulating hematopoietic compartments (15-17). PIG-A- and normal cells in general behave similarly in paired tissue culture assays. One group reported a major difference between PIG-A- cells and controls in susceptibility to apoptosis (18), but this finding has not been reproduced in other laboratories (19) (including our own; unpublished data) (Notably, granulocytes from patients with a variety of marrow failure syndromes, including PNH, have been described as globally resistant to apoptosis (20).) Additionally, there is no known molecular mechanism by which GPI-anchored proteins would influence apoptotic cell death. Marrow cells from some PNH patients showed superior survival in SCID mice (20), although the normal cells in these experiments showed unusually poor persistence, and the mice retain some immunological function. In our recent analysis of the transcriptome of paired normal and PNH cells, including cell lines (K562, Ramos) and immortalized lymphoblastoid cells derived from patients, using DNA chip technology to detect differences in gene expression, there were remarkably few differences, and none were consistent (our unpublished data). These results suggest that the PIG-A mutation is not associated with major changes in the cell's program for growth, differentiation, or death. (21)

In a second model, the *PIG-A* defect must be combined with some extrinsic factor to promote clonal expansion. Genetic analysis of *PIG-A* mutations has indicated that patients often have multiple PNH clones, which may persist for many years (22), consistent with the emergence of clones under selective pressure. Lymphocytes of PNH phenotype and *PIG-A* genotype have appeared in lymphoma patients treated with Campath-1, a monoclonal antibody that coincidentally recognizes a GPI-anchored protein (21) The frequent association of PNH with aplastic anemia first suggested that the hypocellular marrow milieu might be such a permissive environment (23; 24). However, while PNH clones are present in a very large proportion of aplastic anemia patients, PNH does not occur after chemotherapy or radiation-induced aplasia, following stem cell transplant, or in constitutional Fanconi anemia (25). The presence of PNH does appear to correlate in particular with immunologically mediated marrow failure, and in our experience serves as a positive predictor of response to immunosuppresive therapy in aplastic anemia and myelodysplasia (odds ratio of 1.9 for both groups) (26). Furthermore, the presence of PNH clonal expansion, rather than aplastic anemia, is itself strongly associated with the HLA.-DR2 class II histocompatability locus (26).

In most patients with acquired aplastic anemia, the underlying pathophysiology appears to be immune-mediated (27). In vitro inhibition of hematopoiesis by lymphocytes or T cells was an early clue to this mechanism (28); many of the patients tested probably also had expanded PNH clones, and T cell suppression of erythroid cultures also has been described for one PNH patient without marrow aplasia (29). Patients with aplastic anemia frequently show evidence of lymphocyte activation, especially of cytotoxic T cells, which overexpress type I cytokines like interferon and tumor necrosis factor which in turn induce Fas-mediated apoptosis in hematopoietic target cells (27). While the antigens that trigger autoimmunity are unknown, fine analysis of the T cell receptor repertoire has suggested that the immune response is indeed antigen -driven in both aplastic anemia (30) and PNH (31). In our recent study of patients with PNH/aplasia who shared class II HLA.-DR2, we found not only surprisingly limited utilization of the Vβ chain of the T cell receptor, but that among different patients the dominant CD4 T cell clones used an identical CDR3 region for antigen binding. (32).

One possible mechanism for clonal expansion in PNH is escape from immune surveillance, as a number of mammalian GPI-anchored proteins are ligands for T cell receptors, including CD58 and CD59 (for CD2), or serve immune functions, as do CD14 and the FcIII receptor; the absence of such a molecule might result in a global failure of PNH cells to be recognized by the immune system (immune escape is a proposed role of protozoan GPI-anchored proteins like variable surface glycoprotein of trypanosomes). For one inhibitory T cell clone derived from an aplastic anemia patient, anti-CD59 antibody appeared to block cytotoxicity against lymphoblastoid cell targets (33). We described a GPI-anchor protein-deficient

Ramos cell line that was extremely resistant to natural killer cell lysis, but the *PIG-A* gene in these cells was normal (34). Recently, an interesting mechanism was described for proinflammatory cytokine production after heat shock protein stimulation through CD14 (35). Nevertheless, extensive efforts in several laboratories to determine general differences between normal and *PIG-A*- target cells in a wide variety of immunologic assays have been negative (36; 36) (and others unreported). In our unpublished studies, for example, using paired normal and *PIG-A*- cell lines and primary clinical samples as targets, we have been unable to document differences in susceptibility to autologous and allogeneic T cell cytotoxicity or ability to activate T cells, or in their proliferation after challenge with cytokines or Fas agonists. Finally, most clinical data do not show changes that might be predicted in either the presence or size of the PNH clone, before and after immunosuppresive therapy, or on hematologic recovery and at recurrence of pancytopenia; in most cases, the proportion of GPI-anchored protein-deficient cells remains relatively constant over time (10).

In recently published work from Case Western Reserve University, a major difference was observed in cell culture of PNH patients' marrows, but as a deficiency among the "normal" cells rather than an advantage for the *PIG-A* mutant cells: CD34 cells, separated based on the presence or absence of CD59, were cultivated in the presence of growth factors; the cells of PNH phenotype behaved normally, producing several hundred thousand progeny, while CD34 cells of normal phenotype (but from the PNH patients) showed little growth. An inference from these results was that the PNH cells had survived an in vivo injury sustained by the normal progenitors. That this insult might indeed be immunologic was suggested by the expression of Fas on the normal but not on the GPI-anchor protein deficient cells and concomitant sensitivity to Fas-ligand mediated apoptosis (37). We have broadly reproduced these results in a larger number of patients: compared to CD34 cells of the PNH phenotype, normal cells form the patients' bone marrow show lower cloning efficiency in semisolid medium, increased CD95 expression, and frank evidence of apoptosis mainly in the cells of normal phenotype as determined by annexin V and caspase 3 staining (our unpublished data). Longer term culture of these marrow specimens suggest that once dying cells have progressed through apoptosis, the remaining viable cells of normal and PNH phenotype have similar growth and survival characteristics – consistent with selective injury in vivo rather than an intrinsic cellular difference.

Nevertheless, the character of the interaction of a *PIG-A*- clone and either the normal hematopoietic environment or the failed marrow milieu —and indeed of the cause and effect sequence of PNH and aplastic anemia–remains frustratingly enigmatic. Several relationships can be hypothesized, and tested. In the first, a general deficit in effector recognition of a defective cellular target, as for example due to decreased binding of costimulatory molecules secondary to the GPI-anchor protein deficiency, would explain the emergence of normally quiescent but now resistant PNH clones under conditions of autoimmune attack. Most experiments that directly test this possibility have been negative; nevertheless, the effect could be subtle and only manifest under special in vivo circumstances. Dependence of PNH cell viability on the presence of immune mediators is hard to reconcile with the stability of PNH clones over time and despite changes in blood counts and treatment regimens. Alternatively, the mutated PNH cell might have a more primary role in the development of bone marrow failure. GPI-anchored proteins might serve as antigens, either due to molecular mimicry with exogenous peptides (perhaps derived from a viral protein) or subsequently involved as part of the process of antigenic spread. The PNH clone itself might induce autoimmunity. One possibility that we are currently investigating is the consequence of a global alteration in protein trafficking due to absence of GPI anchors. Essentially, what is the fate of GPI-anchorless proteins in the PNH cell? Under normal circumstances, membrane-bound protein degradation should occur largely via the lysosome, leading to presentation of peptides in the context of class II HLA. molecules. In contrast, GPI-anchorless proteins in the *PIG-A-* cell should be processed by the proteasome and their peptides displayed by class I HLA molecules. Published experiments indicate that for chimeric genes transfected into cell lines, constructs with defective carboxyl termini produced proteins which were retained in the endoplasmic reticulum (38) and degraded via the proteasome (39). A model in which protein trafficking and degradation were primary to the pathophysiology of PNH would provide a role for both GPI-anchored protein antigenicity (increased or novel peptide presentation by class I MHC) and a mechanism for escape (decreased class II display). Under normal circumstances, the PNH clone would be suppressed (as are presumably other genetically altered hematopoietic cells, like cells bearing the bcr-abl translocation which can be detected by gene amplification methods in many normal individuals' bone marrow); in other, "tolerant" individuals, clonal expansion might occur unimpeded leading to predominantly hemolytic PNH; in the setting of an aberrant or exaggerated immune response, normal hematopoiesis would suffer, leading to aplastic anemia. There may be other major, unanticipated consequences of failed GPI anchor synthesis beyond deficient expression of certain membrane proteins.

In PNH, models and experiments will need to account for a number of remaining puzzling questions, both clinically relevant and of general biologic importance. Biochemically, little is understood of the presumed special function of the GPI-anchor in the mammalian cell: in altering the mobility of membrane proteins, in internal signal transduction, in collaboration with other proteins in rafts, and in cell-cell protein transfer, and even for shedding from the cell surface (there is sufficient inactive phospholipase C in plasma to release virtually all GPI-anchor proteins from circulating cells). For stem cell biology, why is there such a tiny proportion of *PIG-A-* progeny in normal individuals; why do *PIG-A-* clones persist for years, even decades, supplying the entire blood production of a patient; and why do clones appear to extinguish entirely in rare individuals? Clinically, we need to account not only for the close relationship of a clonal, somatically acquired genetic disorder and immunologically mediated bone marrow failure, but also for evidence of generalized marrow depression in all forms of PNH, the multiplicity of clones in disease, and the progression to refractory fatal thrombosis or aplastic anemia in many patients.

References

1. Rosse, W (2000) A Brief History of PNH. In *PNH and the GPI-linked Proteins*, ed. NS Young, J Moss. 1-20. Academic Press, San Diego, CA.
2. Frickhofen N, Rosenfeld SJ (2000) Immunosuppressive treatment of aplastic anemia with antithymocyte globuilin and cyclosporine. Semin Hematol 37:56-68
3. Luzzatto L, Nafa K (2000) Genetics of PNH. In *PNH and the GPI-linked Proteins*, ed. NS Young, J Moss. 21-48. Academic Press, San Diego, CA.
4. Okazaki I, Moss J (2000) The Function of GPI-Anchored Proteins. In *PNH and the GPI-linked Proteins*, ed. NS Young , J Moss. 159-177. Academic Press, San Diego, CA.
5. Schofield, L (2000) GPI in Lower Animals. In *PNH and the GPI-linked Proteins*, ed. NS Young, J Moss. 179-198. Academic Press, San Diego, CA.
6. Sevlever D, Chen R, Medof ME (2000) Synthesis of the GPI Anchor. In *PNH and the GPI-linked Proteins*, ed. NS Young, J Moss. 199-220. Academic Press, San Diego, CA.
7. Takeda J, Miyata T, Kawagoe K, Iida Y, Endo Y, Fujita T, Takahashi M, Kitani T, Kinoshita T (1993) Deficiency of the GPI anchor caused by a somatic mutation of the PIG-A gene in paroxysmal nocturnal hemoglobinuria. Cell 73:703-11
8. Oni SB, Osunkoya BO , Luzzatto L (1970) Paroxysmal noctural hemoglobinuria: evidence for monoclonal origin of abnormal red cells. Blood 36:145-52
9. Schrenzenmeier H, Hertenstein B, Wagner B, Raghavachar A, Heimpel H (1995) A pathogenetic link between aplastic anemia and paroxysmal nocturnal hemoglobinuria is suggested by a high frequency of aplastic anemia with a deficiency of phosphatidylinositol glycan proteins. Exp Hematol 23:81-7
10. Maciejewski JP, Rivera C, Dunn D, Young NS (2001) Clinical features of the relationship between aplastic anemia and paroxysmal noctural hemoglobinuria. Brit J Haem . In Press
11. Rotoli B, Robledo R, Luzzatto L (1982) Decreased number of circulating BFU-Es in paroxysmal nocturnal hemoglobinuria. Blood 60:157-9
12. Moore JG, Humphries RK, Frank MM, Young N (1986) Characterization of the hematopoietic defect in paroxysmal nocturnal hemoglobinuria. Exp Hematol 14:222-9
13. Issaragrisil S, Piankijagum A, Chinprasertsuk S, Kruatrachue M (1986) Growth of mixed erythroid-granulocytic colonies in culture derived from bone marrow of patients with paroxysmal nocturnal hemoglobinuria without addition of exogenous stimulator. Exp Hematol 14:861-6
14. Araten DJ, Nafa K, Pakdeesuwan K, Luzzatto L (1999) Clonal populations of hematopoietic cells with paroxysmal nocturnal hemoglobinuria genotype and phenotype are present in normal individuals. Proc Natl Acad Sci U S A 96:5209-14

15. Kinoshita T, Bessler M, Takeda J (2000) Animal Models of PNH. In *PNH and the GPI-linked Proteins*, ed. NS Young, J Moss. 139-158. Academic Press, San Diego, CA.
16. Dunn D, Yu J, Nagarjan S, Devetten M, Weichold F, Medof E, Young NS , Liu J (1996) A knock-out model of paroxysmal nocturnal hemoglobinuria: PIG-A- hematopoiesis is reconstituted following intercellular transfer of GPI-anchored proteins. Proc Natl Acad Sci U S A 93:7938-43
17. Kawagoe K, Kitamura D, Okabe M, Taniuchi I, Ikawa M, Watanabe T, Kinoshita T, Takeda J (1996) Glycosylphosphatidylinositol-anchor-deficient mice: implications for clonal dominance of mutant cells in paroxysmal nocturnal hemoglobinuria. Blood 87:3600-6
18. Brodsky RA, Vala MS, Barber JP, Jones RJ, Medof ME (1997) Resistance to apoptosis caused by PIG-A gene mutations paroxysmal nocturnal hemoglobinuria. Proc Natl Acad Sci U S A 94:8756-60
19. Ware RE, Nishimura J, Moody MA, Smith C, Rosse WF, Howard TA (1998) The PIG-A mutation and absence of glycosylphosphatidylinositol-linked proteins do not confer resistance to apoptosis in paroxysmal nocturnal hemoglobinuria. Blood 92:2541-50
20. Iwamoto N, Kawaguchi T, Horikawa K, Nagakura S, Kagimoto T, Suda T, Takatsuki K, Nakakuma H (1996) Preferential hematopoiesis by paroxysmal nocturnal hemoglobinuria clone engrafted in SCID mice. Blood 87:4944-8
21. Hertenstein B, Wagner B, Bunjes D, Duncker C, Raghavachar A, Arnold R, Heimpel H, Schrezenmeier H (1995) Emergence of CD52-, phosphatidylinositolglycan-anchor-deficient T lymphocytes after in vivo application of Campath-1H for refractory B-cell non-Hodgkin lymphoma. Blood 86:1487-92
22. Bessler M, Mason PJ, Hillmen P, Miyata T, Yamada N, Takeda J, Luzzatto L, Kinoshita T (1994) Paroxysmal nocturnal haemoglobinuria (PNH) is caused by somatic mutations in the PIG-A gene. EMBO J 13:110-7
23. Young NS (1992) The problem of clonality in aplastic anemia. Dr. Dameshek's riddle, restated. Blood 79:1385-92
24. Luzzatto L, Bessler M, Rotoli B (1997) Somatic mutations in paroxysmal nocturnal hemoglobinuria: a blessing in disguise. Cell 88:1-4
25. Dunn DE, Tannawattanacharoen P, Boccuni P, Nagakura S, Green SW, Kirby MR, Swanson SJ, Kumar MSA, Rosenfeld S, Young NS (1999) Paroxysmal nocturnal hemoglobinuria cells in patients with bone marrow failure syndromes. Ann Intern Med 131:401-8
26. Maciejewski JP, Follman D, Rivera CE, Brown K, Simonis T, Young NS (2001) Increased frequency of HLA-DR2 in patients with paroxysmal nocturnal hemoglobinuria and PNH/aplastic anemia syndrome. Blood. In Press
27. Young NS and Maciejewski J (1997) The pathophysiology of acquired aplastic anemia. N Engl J Med 336:1365-72
28. Young NS. (1994) Pathophysiology II: immune suppression of hematopoiesis. In *Aplastic Anemia, Acquired and Inherited*, ed. NS Young, BP Alter. 68-99. W.B. Saunders, Philadelphia.

29. Merchav S, Tatarsky I, Chezar J, Sharon R, Rosenbaum H, Schechter Y (2000) Paroxysmal nocturnal hemoglobinuria associated with in vitro inhibition of erythropoiesis by bone marrow T lymphocytes. IMAJ 2:22-4
30. Zeng W, Nakao S, Takamatsu H, Yachie A, Takami A, Kondo Y, Sgimori N , Yamazaki H, Miura Y, Shiobara S, Matsuda T (1999) Characterization of T-cell repertoire of the bone marrow in immune-mediated aplastic anemia: Evidence for the involvement of antigen-driven T-cell response in cyclosporine-dependent aplastic anemia. Blood 93:3008-16
31. Karadimitris A, Manavian JS, Thaler HT, Notaro R, Araten DJ, Nafa K, Roberts IAG, Weksler ME, Luzzatto L (2000) Abnormal T-cell repertoire is consistent with immune process underlying the pathogenesis of paroxysmal nocturnal hemoglobinuria. Blood 96:2613-20
32. Zeng,W, Maciejewski,JP, Chen,G, Young,NS (2001) Limited heterogeneity of T-cell receptor VB usage in aplastic anemia. Journal of Clinical Investigation 108:765-773
33. Takami A, Zeng W, Wang H, Matsuda T, Nakao S (1999) Cytotoxicity against lymphoblastoid cells mediated by a T-cell clone from an aplastic anemia patient: role of CD59 on target cells. Br J Haematol 107:791-6
34. Dunn DE, Nagakura S, Tanawattanacharoen P, Knez J, Medof ME, Young NS (1997) The PNH phenotype (GPI-anchored protein-deficiency) protects hematopoietic targets against lymphotcytoxic attack. *Blood* 90(Supplement 1) :407a (Abstr.)
35. Asea A, Kraeft SK, Kurt-Jones EA, Stevenson MA, Chen LB, Finberg RW, Koo GC, Calderwood SK (2000) HSP70 stimulates cytokine production through a CD14-dependent pathway, demonstrating its dual role as a chaperone and cytokine. Nat Med 6:435-42
36. Hollander N, Selvaraj P, Springer TA (1988) Biosynthesis and function of LFA-3 in human mutant cells deficient in phosphatidylinositol-anchored proteins. J Immunol 141:4283-90
37. Chen R, Nagarajan S, Prince GM, Maheshwari U, Terstappen LW, Kaplan DR, Gerson SL, Albert JM, Dunn DE, Lazarussw HM, Medof ME (2000) Impaired growth and elevated FAS receptor expression in PIGA(+) stem cells in primary paroxysmal nocturnal hemoglobinuria. J Clin Invest 106:689-96
38. Fields MC, Morani P, Li W, Keller G-A, Caras IW (1994) Retention and degradation of proteins containing an uncleaved glycosylphosphatidylinosital signal. J Biol Chem 269:10830-7
39. Oda K, Ikehara Y, Omura S (1996) Lactacystin, an inhibitor of the proteasome, blocks the degradation of a mutant precusor of glycosylphosphatidylinositol-linked protein in a pre-Golgi compartment. Biochem Biophys Res Commun 219:800-5

Part 6

PNH, Aplastic Anemia, Myelodysplastic Syndromes (1): Pathology

Does PNH Ever Arise de Novo?

E.C. Gordon-Smith

Department of Haematology, St George's Hospital Medical School, Cranmer Terrace, London SW17 0RE, UK

Summary: PNH occurs in a spectrum of clinical settings from haemolytic PNH with cellular marrow, haemoglobinuria, recurrent abdominal pain and a tendency to thrombosis, to a laboratory finding on the background of aplastic anaemia where the bone marrow failure dominates the clinical syndrome. The frequent association with aplastic anaemia promoted the hypothesis that the progeny of the PIG-A mutated haemopoietic stem cells (HSC) had a growth advantage and escaped from the inhibition of the normal HSC in that setting. The hypothesis is supported by the findings of PNH clones in normal individuals, clones which do not become clinically significant. The question explored in this article is whether PNH clones ever have a proliferative advantage over normal marrow, and hence present a clinically significant disease without marrow suppression. Whilst not entirely conclusive, evidence suggests that there is always marrow depression even in haemolytic PNH. The question of whether this marrow suppression is similar to that seen in AA is discussed.

Key words: Paroxysmal nocturnal haemoglobinuria, Bone marrow failure, Aplastic anaemia

PNH used to be a clinical disease with readily recognisable features – haemoglobinuria occurring typically after sleep, haemolytic anaemia, abdominal pain, a risk of thrombophlebitis and thromboembolism accompanied by haemosiderinuria. These were the classic symptoms of a readily recognised disorder which had a number of eponyms and synonyms [1]. The introduction of the acidified serum lysis test in 1938 allowed the disorder to be distinguished from other causes of acute acquired intravascular haemolysis. The occurrence of haemoglobinuria, apparently promoted by sleep, but not influenced by cold, also distinguished the disease from other causes. The disease was extensively studied by Dacie, who wrote in 1967, 'The most likely mechanisms for PNH seems to the author to be the development of self-perpetuating change, possibly the result of somatic mutation, which affects the haemopoietic stem cells of the marrow and which leads to .... defective erythrocytes, leukocytes and platelets' [1]. The association with aplastic anaemia, at that time a feared disease with a high mortality, was also recognised as a remarkable association [2]. PNH clones arising on the background of aplastic anaemia did not necessarily lead to a significant haemolytic anaemia. It became customary, at least in Dacie's unit, to talk about haemolytic PNH and 'laboratory PNH' – the latter when the Ham's test was positive but there was no evidence of significant haemoglobinuria. The course of classical PNH was slow and somewhat unpredictable – some patients losing the PNH cells after many years, others succumbing to thrombosis or, more rarely, to the results of pancytopenia. Subsequent follow-up of Dacie's patients [3] in the UK [4] and in the study of Socié and colleagues in France [5] showed that the median survival of patients with haemolytic PNH was 10-15 years, with many patients surviving with stable disease beyond that. This of course was in stark contrast to the poor survival of patients with aplastic anaemia, which in the same time period had a median survival measured in months rather than years [6,7].

The association with aplastic anaemia gave rise to the notion that the PNH cells, which we now of course know lack the GPI-anchored proteins, had a growth advantage over the damaged aplastic stem cells, perhaps by evading immune attack. Candidate antigens either in the GPI-anchored surface proteins or the GPI anchor itself have been extensively sought as possible triggers of autoimmune antigen suppression, so far without any being definitely identified. The corollary of the growth advantage theory was that if clones of PNH cells arose in normal marrow they would not have the growth advantage and hence would not be phenotypically expressed or become clinically significant, the so-called escape theory [8].

The discovery of the molecular basis of PNH [9] and the development of immune phenotyping for GPI anchored proteins, allowed the identification of very small populations of GPI protein-negative cells in a population, and confirmed that such negative cells did indeed carry the mutation [10]. As we have also heard at this Symposium, a very large number of mutations of the PIG-A gene have been described which lead to the phenotype of the PNH clone with either complete or partial loss of synthesis of the GPI anchor. Furthermore, multiple distinct clones due to different PIG-A mutations frequently occur both in haemolytic PNH and in AA/PNH syndrome [11,12]. Thus there appears to be a situation in which there is frequent somatic mutation of the PIG-A gene within the normal marrow, but that the consequences of this mutation are not apparent because of the normal growth of haemopoietic stem cells. This theory supports the idea that the PNH clone is only experienced clinically when there is a growth advantage – as in AA – and explains the source of the clone from the frequently mutating clones within the bone marrow. However, it does not explain the observation that PNH clones tend to be stable, nor the spontaneous appearance of haemolytic PNH unaccompanied by gross marrow failure or other haematological disorder. There are several questions to be answered in considering haemolytic PNH. In haemolytic PNH, what allows the clone to be expressed despite apparent normal cellularity? Is the proliferation and differentiation of the non-PNH marrow normal? If there is a background suppression of normal haemopoiesis, what is the mechanism? If aplastic anaemia is an autoimmune disease, do the same processes play a part in haemolytic PNH acting on the non-mutated marrow? If haemolytic PNH always occurs in the context of suppression of normal haemopoiesis, are the PNH clones opportunistic by-standers or are they in some way responsible for the suppression of normal haemopoiesis?

Stem Cells in PNH

Marrow function in vitro, even in haemolytic PNH with cellular marrow, is depressed. Circulating BFU-E are decreased compared with numbers [13]. The total number of CD34 positive cells is reduced several fold compared with normal, the reduction applying both to CD59 positive and CD59 negative haematopoietic stem cells (HSC) [14]. These and other studies confirmed that the CD59 negative, CD34 positive HSC had a poorer proliferative potential than normal HSC. Chen and colleagues separated out the CD59 positive and CD59 negative cells from the same individual with PNH and showed that the mutated clones, when grown in liquid culture, exhibited growth and differential patterns similar to those of CD34 positive cells from normal controls, but that the non-PIG-A mutated cells in the same individuals had markedly impaired growth, and furthermore found that the Fas receptor on the non-mutated cells was elevated [15]. In these studies the CD34 positive cells are sorted to a high degree of purity and the culture therefore takes place in the absence of T cells or other accessory marrow cells. It would appear then that in PNH clinical manifestations are always accompanied by suppression of normal bone marrow function which allows the 'escape' of the PNH cells [8]. This still begs the question of whether the PNH clone (or clones) is responsible for the suppression of the non-mutated stem cell or whether an extrinsic cause is responsible.

Immune Processes in PNH and Aplastic Anaemia

It is generally felt that autoimmune processes, probably through the T cell cytotoxic repertoire, underlie the pathogenesis of AA. The subject has been extensively reviewed [16,17]. So far, antigenic determinants which might have provoked a T cell response, have not been identified in normal haemopoietic cells, but there has been some evidence that there may be over-expression of certain Vβ families in the lymphocyte repertoire in the bone marrow in some AA patients, suggesting an oligoclonal proliferation [18]. Whether there are specific Vβ clones which are likely to arise in aplastic anaemia is not known, nor whether this is specific for AA/PNH, or a consequence of the bone marrow depression, is also unclear. Japanese patients with AA, who had the HLA DRB1*1501 allele, showed expansion of certain T cell clones from the bone marrow of the β variable chain genes [19]. Karadimitris and colleagues examined the T cell repertoire in patients with haemolytic PNH. They found an increased frequency of clonal T cell expansion in a proportion of the PNH patients and such expansions persisted long term in some. The size of the expansion was similar to that which had been found in aplastic anaemia patients [20]. On the other hand they were unable to demonstrate any difference in sensitivity of normal and PNH EBV transformed B cell lymphoblastoid cell lines to the cytotoxic effects of autologous EBV specific T cell lines and clones from a patient with PNH [21].

There are thus intriguing similarities between aplastic anaemia, AA/PNH and haemolytic PNH in the immune disturbances which can be identified, most notably oligoclonal expansions in the T cell repertoire. This might be interpreted to mean that the underlying processes which allow the emergence of the PNH clones are the same, or at least similar, in both AA, PNH and in haemolytic PNH, with background inhibition of normal haemopoiesis. The size of the PNH clone would then be related to the proliferative potential of that particular mutation, in the context of the damaged marrow.

Implications for Treatment

If immune suppression of normal haemopoiesis allows the emergence of the PNH clone, then treatment or interruption of the immune process, as with AA alone, should allow at least partial recovery of normal haemopoiesis and hence loss of the supposed growth advantage of the PNH clone. The PNH clone should therefore diminish in clinical severity. Certainly patients with haemolytic PNH have been treated with antilymphocyte globulin or cyclosporin and improvements reported [22]. In most instances those patients who have most evidence of bone marrow failure have the better response to antilymphocyte globulin – as perhaps would be expected. Patients with haemolytic PNH on the other hand have been reported anecdotally as showing some improvement in overall haemopoiesis. At the other end of the spectrum, patients with AA/PNH have the same response to immunosuppression as AA patients without a PNH clone [23]. The differences in response may not be absolute, but differences in degree, related to the drive for the immune suppression of haemopoiesis. The relatively poor results of allogenic HSC transplants in PNH [24] might be related to the thrombotic tendency rather than any intrinsic problems with the marrow.

Discussion

The simple days when PNH was characterised by its clinical presentation and the presence of PNH clones in aplastic anaemia was an interesting distraction are over. It is now recognised that the PNH mutation is common and occurs in normal marrow and that probably when there is a significant expansion of the PNH clone there is always suppression of the unaffected haemopoiesis. The expansion of the PNH clone has been described in Fanconi anaemia, clearly a pre-existing bone marrow depression not produced by an immune process. In this extraordinary instance, the disappearance of the PNH clone was accompanied by restoration of normal haemopoiesis in the previously severely affected Fanconi anaemia [25,26]. However, this remarkable case does not vitiate the notion that generally there is an acquired suppression of haemopoietic function which allows the PNH clone to expand. It is possible that the presence of a PNH clone and its identification by the immune system as 'self' actually leads to the immune suppression of normal haemopoietic function in which case PNH could be considered a primary event.

References

1. Dacie JV (1967) Paroxysmal nocturnal haemoglobinuria. In: The Haemolytic Anaemias, Congenital and Acquired, Part 4. Drug induced haemolytic anaemias, etc. 2nd edn, Churchill, London, pp 1128-1260
2. Dacie JV, Lewis SM (1961) Paroxysmal nocturnal haemoglobinuria: variation in clinical severity and association with bone marrow hypoplasia. Brit J Haematol 7:442-457
3. Dacie JV, Lewis SM (1972) Paroxysmal nocturnal haemoglobinuria: clinical manifestations, haematology and nature of the disease. Ser Haematol 5:3-23
4. Hillmen P, Lewis SM, Bessler M, et al (1995) Natural history of paroxysmal nocturnal hemoglobinuria. New Eng J Med 333: 1253-1258
5. Socié G, Mary J-Y, de Gramont A, et al (1996) Paroxysmal nocturnal haemoglobinuria: long term follow up and prognostic factors. Lancet 348:573-577
6. Williams DM, Lynch RE, Cartwright GE (1973) Drug-induced aplastic anaemia. Semin Hematol 10:195-223
7. Camitta BM, Rappeport TM, Parkman R et al, (1975) Selection of patients for bone marrow transplantation in severe aplastic anaemia. Blood 45:355-363
8. Rotoli B, Luzzatto L (1989) Paroxysmal nocturnal haemoglobinuria. Baillières Clin Haematol 2:113-138
9. Takeda J, Miyata T, Kawagoe K, et al (1993) Deficiency of the GPI anchor caused by a somatic mutation of the PIG-A gene in paroxysmal nocturnal hemoglobinuria. Cell 73:703-711
10. Araten DJ, Nafa K, Pakdeesuwan K, et al (1999) Clonal population of hematopoietic cells with paroxysmal nocturnal hemoglobinuria genotype and phenotype are present in normal individuals. Proc Natl Acad Sci USA 96:5209-5214
11. Nishimura J, Inoue N, Wada H (1997) A patient with paroxysmal nocturnal hemoglobinuria bearing four independent PIG-A mutant clones. Blood 89:3470-3476
12. Bessler M, Mason P, Hillmen P, et al (1994) Somatic mutations and cellular selection in paroxysmal nocturnal haemoglobinuria. Lancet 343:951-953
13. Rotoli B, Robledo R, Luzzatto L (1982) Decreased number of circulating BFU-E in paroxysmal nocturnal hemoglobinuria. Blood 60:157-159
14. Maciejewski JP, Sloand EM, Sato T, et al (1997) Impaired haematopoiesis in paroxysmal nocturnal haemoglobinuria/aplastic anaemia is not associated with a selective proliferative defect in the glycosylphosphatidylinositol-anchored protein-deficient clone. Blood 89:1173-1181
15. Chen R, Nagarajan S, Prince GM (2000) Impaired growth and elevated Fas receptor expression in PIG-A positive stem cells in primary paroxysmal nocturnal haemoglobinuria. J Clin Invest 106:689-696
16. Young NS (1994) Pathophysiology II: Immune suppression of haematopoiesis. In: Aplastic Anaemia, Acquired and Inherited. Eds. Young NS, Alter BP. Sanders, Philadelphia, pp. 68-99
17. Nakao S (2000) Role of T lymphocytes in the pathophysiology of aplastic anaemia. In: Aplastic Anaemia Pathophysiology and Treatment. Eds. Schrezenmeier H, Baccigalupo A. Cambridge University Press, Cambridge, pp. 41-57
18. Manz CY, Dietrich PY, Schnuriger V, et al (1997) T cell receptor β-chain variability in bone marrow and peripheral blood in severe acquired aplastic anaemia. Blood Cells Mol Dis 23:110-122

19. Zeng W, Nakao S, Takamatsu H, et al (1999) Characterisation of T-cell repertoire of the bone marrow in immune-mediated aplastic aaemia; evidence for the involvement of antigen-driven T cell response in cyclosporin-dependent aplastic anemia. Blood 93: 3008-3016
20. Karadimitris A, Manavalan JS, Thaler HT, et al (2000) Abnormal T cell repertoire is consistent with the immune process underlying the pathogenesis of paroxysmal nocturnal hemoglobinuria. Blood 96:2613-2620
21. Karadimitris A, Notaro R, Koehne G, et al (2000) PNH cells are as sensitive to T cell-mediated lysis as their normal counterparts: implications for the pathogenesis of paroxysmal nocturnal haemoglobinuria. Brit J Haematol 111:1158-1163
22. Paquette RL, Yoshimura R, Veiseh C, et al (1997) Clinical characteristics predict response to antithymocyte globulin in paroxysmal nocturnal haemoglobinuria. Brit J Haematol 96:92-97
23. De Lord C, Tooze JA, Saso R, et al (1998) Deficiency of glycosylphosphatidyl-inositol-anchored proteins in patients with aplastic anaemia does not affect response to immunosuppressive therapy. Brit J Haematol 101:90-93
24. Saso R, Marsh J, Cevreska L, et al (1999) Bone marrow transplants for paroxysmal nocturnal haemoglobinuria. Brit J Haematol 104:392-396
25. Dacie JV, Gilpin A (1944) Refractory anaemia (Fanconi type). Arch Dis Child 19:155-162.
26. Dacie JV, Lewis SM (1961) Paroxysmal nocturnal haemoglobinuria: variations in clinical severity and association with bone marrow hypoplasia. Brit J Haematol 7:442-457

Telomeres in Myelodysplastic Syndrome and Its Related Disorders: Does Telomere Length Reflect Stem Cell Turnover in Clonal Hematopoietic Disorders?

Junko H. Ohyashiki[1,2], Goro Sashida[2], Tsutomu Shichishima[3], and Kazuma Ohyashiki[2]

[1]Department of Virology, Medical Research Institute, Tokyo Medical and Dental University, 1-5-45, Yushima, Bunkyo-ku, Tokyo 113-8510, Japan
[2]The First Department of Internal Medicine, Tokyo Medical University, 6-7-1, Nishishinjuku, Shinjuku-ku, Tokyo 160-0023, Japan
[3]The First Department of Internal Medicine, Fukushima Medical University, Hikarigaoka, Fukushima, 960-1295, Japan

Summary: Normal hematopoietic cells express telomerase activity, however the presence of telomerase does not necessarily imply stable and thus unchanging telomere length. Age-adjusted telomere length was lower in patients with both refractory anemia (a low grade subtype of myelodysplasric syndrome: MDS) and aplastic anemia. Telomere length may reflect both the cellular proliferative history and genomic instability in MDS, while telomere shortening in aplastic anemia might depend on the severity of stem cell pool contraction. Telomere length in limited number of patients with paroxysmal nocturnal hemoglobinuria (PNH) was highly variable. Gradual telomere loss with rapid cycling of hematopoietic stem cells might contribute to immunosenescence, exhausted hematopoiesis, and increased likelihood of malignant transformation. Therefore, detailed analysis of telomere dynamics in different cell population is needed to clarify the pathogenesis of MDS and related disorders.

Key words: telomere, hematopoietic stem cell, MDS, AA, and PNH

Introduction

Telomere shortening and telomerase in determining cell replicative potential has received considerable attention in the last few years. While the role of telomerase has been extensively evaluated in human cancers, the study of its dysregulation in hematologic disease has been somewhat hindered because normal hematopoietic cells express varying levels of telomerase activity according to their state of differentiation and activation [1-5]. The major concern about telomere dynamics in the field relates to the need for comparison of cells at similar differentiation stages. So far, most of the studies have done using heterogeneous bone marrow cells and/or peripheral blood cells. A commonly used TRF technique (Southern hybridization of terminal restriction fragments) is able to show only the average size of telomeres in heterogeneous cell population, not their differences of individual chromosomes or even a complete loss or extreme expansion at certain chromosome.

To overcome this obstacle, a quantitative analysis of telomere using fluorescence in situ hybridization with peptide nucleic acid (PNA) probes become a tool of choice for detailed analysis of telomeres [6]. In addition, combination of PNA probe and flow cytometry (flow-FISH) method has made it possible to analyze telomere length and thus the replication history of different cell types of cells sorted by FACS [7].

On the basis of these methodological developments, it appears that analysis of telomere length is a reasonable approach to study of replicative history of stem cells in intractable hematologic disease of various etiologies. In this chapter, we discuss about telomere dynamics in myelodysplastic syndrome (MDS), acquired aplastic anemia (AA), and paroxysmal nocturnal hemoglobinuria (PNH).

Telomeres and telomerase in normal hematopoietic cells and their progeny

Normal stem cells of renewal tissues including specific hematopoietic cells expressed some telomerase activity, therefore, we first need to examine recent studies that demonstrate telomerase activity in normal hematopoietic cells, before considering the telomere and telomerase status of hematologic disease.

The most important concept for interpreting the behavior of stem cells is that the presence of telomerase does not necessarily imply stable and thus unchanging telomere length. The initial studies with hematopoietic cells demonstrated that the mean terminal restriction fragments (TRFs) of total nucleated cells from adult bone marrow was shorter than TRFs of the same type of cells isolated from fetal liver and cord blood [8,9]. Short TRFs were found for purified primitive $CD34^{+}CD38^{low}$ cells from adult bone marrow compared to those from fetal liver. In peripheral blood cells, telomere length and telomerase activity progressively decreased with aging [1, 8, 9,11]. These findings support the idea that telomeres also shorten in hematopoietic stem cells (HSCs) during normal replicative aging.

The level of telomerase activity was different in human HSCs and their differentiate progeny [1-4,10,12,13] (Figure 1). In contrast to somatic cells that lack telomerase activity, it is clear that many hematopoietic cells are telomerase-competent. Primitive HSCs are likely to be quiescent most of the time and thus the population of cells exhibits a low level of telomerase activity. However, upon stimulation to proliferate, telomerase activity appears to be upregulated in their immediate progeny and may help to slow down the rate of telomere erosion. The more mature cells then become quiescent again and down-regulate telomerase activity [14]. The telomeres of HSCs shorten, probably due to inadequate levels of telomerase activity that slow but do not prevent telomere erosion.

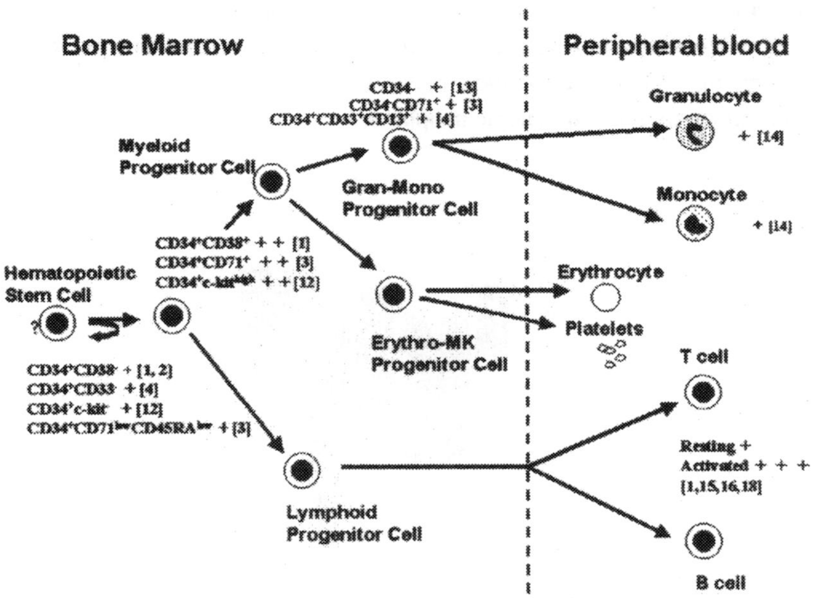

Fig. 1: Telomerase activity in normal, hematopoietic cell populations. **+ - +++:** Level of telomerase activity based on data presented in the reference is also shown.

Both T-cells and monocyte/B cells in peripheral blood also had low telomerase activity which was elevated when either cell types was cultured with mitogen

[1,15-19]. The highest in vivo telomerase expression in normal T cells is present in thymus, followed by T cells in the tonsil [18,19]. The germinal center B-cells also show high telomerase activity [20,21]. Thus, telomerase may play a permissive role in T cell and B cell development and in determining the capacity of lymphoid cells for cell division and clonal expansion.

Telomere shortening associated with disease evolution patterns in myelodysplastic syndrome

MDS is a clonal hemopathy characterized by dysplastic features in blood cells. Since approximately 30% of patients show progression into AML, MDS is known as a preleukemic state [21]. In an earlier report, we have shown that telomere shortening is closely associated with disease evolution patterns [22]. The MDS patients in our study were classified into three groups: The first group had telomere shortening at the time of diagnosis. This group is characterized poor prognosis and complex chromosomal abnormalities. The second group showed reduction of telomere length after leukemic transformation, and the third group showed stable telomere length during the course (Figure 2). This was the first report dealing with the association between complex cytogenetic changes and telomere shortening in hematologic neoplasia.

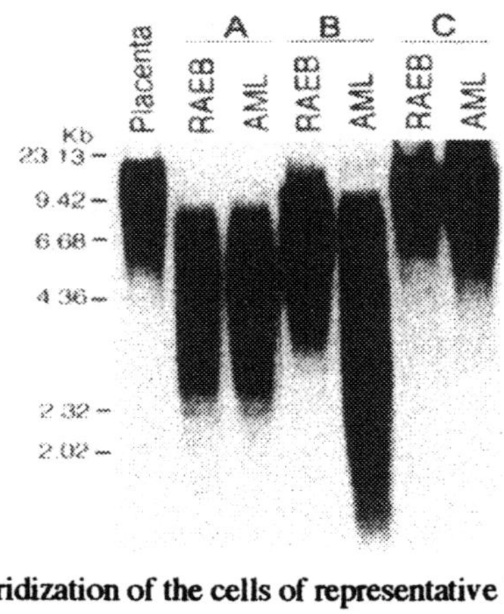

Fig. 2: Southern blot hybridization of the cells of representative MDS patients using 5'-^{32}P-labelled $(TTAGGG)_4$ probe. Changes of telomere length are shown at two different phases; at the time of diagnosis (MDS phase) and when disease had progressed (AML Phase). Pattern A (left): the telomere length at both RAEB phase and AML phase was less than expected. Pattern B (middle): telomere shortened after disease has progressed. Pattern C (right): telomere lengths do not shorten before and after disease progression.

Telomeres and telomerase studies in myelodysplastic syndrome in comparison with *de novo* acute myeloid leukemia

After quantitative system of telomerase activity was established [23], we have found that a high level of telomerase activity is not common in MDS unlike *de novo* acute leukemia, although the level of telomerase activity slightly increases in MDS in comparison with normal bone marrow [24,25]. In fact, variable telomere length has been observed in bone marrow samples with MDS regardless of the percentage of blasts [24,27]. Table 1 summarizes the studies on telomeres and telomerase in MDS in comparison with de novo acute myeloid leukemia [2,4,13,14,22,24-30]. Telomere length was expressed using TRF analysis and telomerase activity was measured using the quantitative telomere repeat amplification protocol (TRAP) assay [23].

Table 1. Telomere length and telomerase activity in myelodysplastic syndrome and acute myeloid leukemia

Diagnosis	TRF (in Kb) Mean ± SD	range	Telomerase activity Normal	Low	High	References
Myelodysplastic syndrome						
	7.6 ± 2.4	3.2~20.1	35.5%	59.1%	5.4%	[2][13]
	(n=165)			(n=93)		[22] [24]-[26]
Acute myeloid leukemia						
	5.2 ± 1.1	2.4~19.6	26.3%	39.2%	41.5%	[2][4][13]
	(n=183)			(n=183)		[14][25]
						[27] -[30]

The references included in this table are shown. Levels of telomerase activity roughly estimated by authors and based on data presented in the references given. high, more than 10-fold compared to normal cells. Low; telomerase activity was elevated compared to normal cells, but less than 10-fold.

Since shortened telomere length might be related to genomic instability and poor prognosis, we measured TRFs in 89 patients with MDS composed of refractory anemia (RA), RA with excess of blasts (RAEB) and RAEB in transformation [29]. Figure 3 clearly shows that patients with shortened telomere has poorer prognosis than those with normal telomere length. The relationship between telomere dynamics and cytogenetic and molecular abnormalities occurring during the progression of MDS remains an intriguing research field, and could contribute to the development of an International Prognostic Scoring System

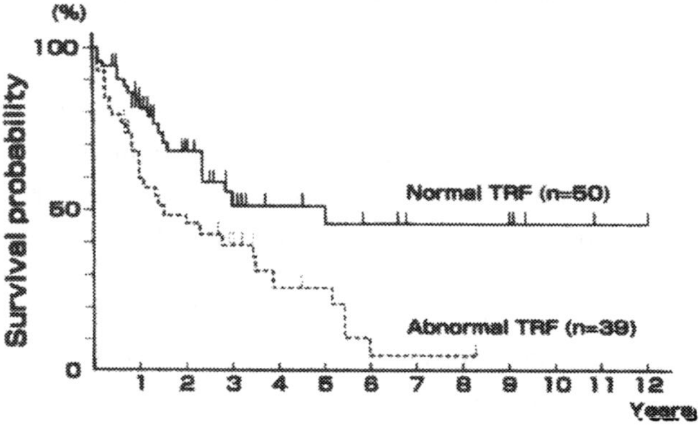

Fig.3 Survival probability of MDS patients separated by peak telomere length. Normal TRF indicates patients within the age-matched normal telomere length. Abnormal telomere length indicates shortened or elongated TRF compared to the age-matched telomere length. A significant difference in survival probability was noted between MDS patients with abnormal TRF and those with normal TRF ($P<0.05$)[24].

(IPSS) for MDS [21]. Therefore, there is increasing interest of telomere length rather than telomerase activity in MDS prognosis.

Telomere dynamics in refractory anemia, aplastic anemia and paroxysmal nocturnal hemoglobinuria

As described above, one would predict the following patterns for telomere length and telomerase activity in hematologic disease: prior to cellular immortalization, telomere length should reflect the proliferative history of the cell (E.G. how far the disease has progressed). Therefore, telomere dynamics in MDS, especially in patients with RA, should be reconsidered from a viewpoint of mitotic history of HSCs. Based on these concepts, we focused TRFs in patients with RA, and then compared TRFs among RA, AA, and PNH. Compared with age-matched normal controls, delta TRF was calculated as reported by other invstigators [31-33].

Telomere length in RA

As shown in Figure 4, the age-adjusted TRF (delta TRF) in patients with total RA was significantly lower than expected.

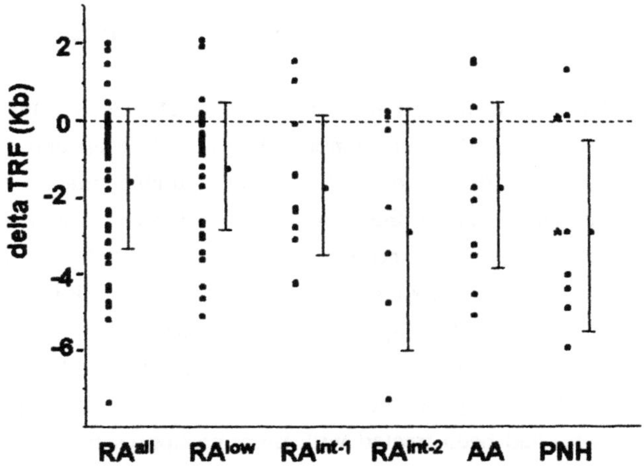

Fig. 4: Age-adjusted telomere length in mononuclear cells from patients with RA, AA and PNH. Patients with RA was separated into 3 groups according to the IPSS for MDS. Astricks indicate secondary PNH.

To further analyze telomere shortening in patients with RA, we subdivided patients with RA on the basis of the severity of the disease according to the IPSS for MDS [21]. According to IPSS for MDS, patients with RA could not be classified into high-risk group, therefore we compared TRFs among patients with low, intermediate-1 and intermediate-2 risk group (Figure 4). There were no significant differences in delta TRF between RA patients with low and intermediate-1 risk group. Strikingly, patients with intermediate-2 risk group showed lower delta TRF compared to those with low and intermediate-1 risk group. These findings confirm our previous investigation dealing with an association between telomere shortening and disease evolution pattern [22]: Shortened telomere at the time of diagnosis (pattern A) might reflect how far disease has progressed in addition to genomic instability, thus relate to aggressive clinical course.

Telomere length in AA and PNH

Average delta TRF in patients with AA was also lower than expected as others reported (Figure 4), suggesting loss at the level of the hematopoietic stem cell [31,33,34]. There was no significant difference in delta TRF between AA and total RA. It is controversial whether telomere length is directly correlated with disease duration in patients with AA [31,33], however telomere shortening may reflect the severity of impairment of hematopoiesis as a result of extensive proliferation of HSCs.

Several lines of evidence suggest that recipients of stem cell transplantation have shorter telomeres than do their donors [35-39]. Allsopp and colleagues reported that telomeres shortening exsit during serial transplantation of murine HSCs [40]. These data suggest that irreversible telomere loss occurs as a consequence of the expansion of normal HSCs in vivo under conditions of replicative stress. Additional telomere loss in such condition might affect the incidence and/or time of onset of age-related hematologic disorders, including secondary malignancies. Therefore, accelerated senescence of HSCs might underline the pathophysiology of secondary MDS in a minor population of patients with AA who had been treated with immunosuppressive agents with or without granulocyte-colony stimulating factor adminstration.

Although delta TRF in PNH was highly variable in the present study (Figure 4), it tended to be lower than expected which was not consistent with the previously reported observations [31,33]. In our study, 5 patients had hemolytic PNH at presentation without evidence for prior history of AA or MDS. By contrast, two patients had had laboratory findins compatible with PNH during the course of MDS (secondary PNH), having been negative at the first manifestation. One patient with secondary PNH was still cytopenic at the time of this study regarding TRF measurement, while the other had fully recoverd blood counts together with an appearance of PNH clone. As GPI-deficient cells arise from a mutation in a single HSC, telomere dynamics in PNH cells might also reflect mitotic history of HSC from which PNH clone came. Comparison of telomere length in PNH and normal clone is needed to clarify the difference of stem cell turnover in these 2 clones.

Aspects of telomere studies in MDS and related disorders

There were several limitations in interpreting telomere dynamics in a series of our TRF studies. First, only a limited number of patients were studied. Second, sequential measurement of delta TRF in the same patient should be performed, as there might be considerable telomere heterogeneity between individuals. Third, analysis of the telomere length in different cell population such as granulocytes and CD34+ cells might be a hopeful approach to understand the stem cell turn over in MDS and its related disorder. Fourth, the intriguing question of different regulation of telomere length will be hopefully answered to find a direct association of telomere shortening in certain chromosomes and disease evolution.

Although there might be a long way to address above-mentioned issues, several methodological approaches have recently been developed. We examined the telomere status in each metaphase using Cy3-labelled Telomere PNA Probe (Telomere PNA FISH Kit, DAKO, Glostrup, Denmark), in order to clarify the telomere status in heterogeneous cell population. We have found that heterogeneity of telomere length in each metaphase in a subset of MDS patients. A mixture of metaphases with shortened telomere signals and with normal telomere signals was seen, even though TRF revealed to be normal telomere length (Sashida et al., unpublished observation, 2000).

Another approaches for detailed analysis of the telomere length in different cell population is combination of PNA probe and flow cytometry (flow-FISH) method. Brummendorf et al. have demonstrated using flow-FISH analysis that the telomeres of granulocytes are shorter in patients with AA than the age-matched telomere lengths, perhaps as a result of rapid turn over of stem cell [34]. Unlike AA, no consistent telomere shortening of granulocytes from patients with RA was found, suggesting genetic heterogeneity of MDS cells (Sashida et al., unpublished observation, 2001). While there are still concerns, these preliminary data suggest the possibility that telomere length might not be the same in cells derived from the same MDS clone. Telo-FISH and flow-FISH analysis will be a helpful approach to clarify a direct connection of telomere shortening and clonal evolution in MDS and its related hematologic disorders.

Futher studies performed with sequentioal samples will help to circumvent some limitations of telomere analysis that are associated with interindividual variation in telomere length. Since some promising areas of tissue engineering include rejuvenation of HSCs for improving stem cell [41-43], these studies are warranted to validate the use telomere length as a measure of mitotic history of HSC in clonal hematopoietic disorders.

Acknowledgements
The authors thank Prof. Jerry W. Shay (University of Texas Southwestern Medical Center) for helpful suggestions, Mr. Takehiro Kasahara (DAKO JAPAN CO., LTD) for supplying flow-fish kit (DAKO Telomere PNA Kit/FITC), and Ms. Ayako Hirota for technical assistance.

References

1. Hiyama K, Hirai Y, Koizumi S, et al (1995) Activation of telomerase in human lymphocytes and hematopoietic progenitor cells. J Immunol 155: 3711-3715

2. Counter CM, Gupta J, Harley CB, et al. (1995) Telomerase activity in normal leukocytes and in hematologic malignancies Blood 85: 2315-2320
3. Chiu CP, Dragowska W, Kim NW, et al (1996) Differential expression of telomerase activity in hematopoietic progenitors from adult human bone marrow. Stem Cells 14: 239-248
4. Zhang W, Pietyszek MA, Kobayashi T, et al (1996) Telomerase activity in human acute myelogenous leukemia: inhibition of telomerase activity by differentiation-inducing agents. Clin Cancer Res 2: 799-803
5. Xu D, Gruber A, Peterson C, et al (1998) Telomerase activity and the expression of telomerase components in acute myelogenous leukaemia. Br J Haematol 102: 1367-1375
6. Lansdorp PM, Verwoerd NP, van de Rijke FM, et al (1996) Heterogeneity in telomere length of human chromosomes. Hum Mol Genet 5: 685-691
7. Rufer N, Dragowska W, Thornbury G, et al (1998) Telomere length dynamics in human lymphocyte subpopulations measured by flow cytometry. Nat Biotechnol 16: 743-747
8. Vaziri H, Dragowska W, Allsopp RC, et al (1994) Evidence for a mitotic clock in human hematopoietic stem cells: loss of telomeric DNA with age. Proc Natl Acad Sci USA 91: 9857-9860
9. Shay JW, Werbin H, Wright WE (1996) Telomeres and telomerase in human leukemias. Leukemia 10: 1255-1261
10. Yui J, Chiu CP, Lansdorp PM (1998). Telomerase activity in candidate stem cells from fetal liver and adult bone marrow. Blood 91: 3255-326
11. Iwama H, Ohyashiki K, Ohyashiki JH, et al (1998) Telomeric length and telomerase activity vary with age in peripheral blood cells obtained from normal individuals. Human Genet 102: 397-402
12. Sakabe H, Yahata N, Kimura T, et al (1998) Human cord blood-derived primitive progenitors are enriched in CD34+c-kit- cells: correlation between long-term culture-initiating cells and telomerase expression. Leukemia 12: 728-734
13. Engelhardt M. Engelhardt M, Kumar R, et al (1997). Telomerase regulation, cell cycle, and telomere stability in primitive hematopoietic cells. Blood 90: 182-193
14. Broccoli D, Young JW, de Lange T (1995) Telomerase activity in normal and malignant hematopoietic cells. Proc Natl Acad Sci USA 92: 9082-9086
15. Norrback KF, Enblad G, Erlanson M, et al (1996). Telomerase activation in normal B lymphocytes and non-Hodgkin's lymphomas. Blood 88: 222-229
16. Igarashi H, Sakaguchi N (1996) Telomerase activity is induced by the stimulation to antigen receptor in human peripheral lymphocytes. Biochem Biophys Res Commun 219: 649-655
17. Igarashi H, Sakaguchi N (1997) Telomerase activity is induced in human peripheral B lymphocytes by the stimulation to antigen receptor. Blood 89: 1299-1307

18. Weng N-P, Levine BL, June CH, et al (1996) Regulated expression of telomerase activity in human T lymphocyte development and activation. J Exp Med 183: 2471-2479
19. Pan C, Xue BH, Ellis TM, et al (1997) Changes in telomerase activity and telomere length during human T lymphocyte senescence. Exp Cell Res 231: 346-353
20. Norrback KF, Enblad G, Erlanson M, et al (1998) Telomerase activity in Hodgkin's disease. Blood 92: 567-573
21. Greenberg P, Cox C, LeBeau MM, et al (1997) International scoring system for evaluating prognosis in myelodysplastic syndromes. Blood 89: 2079-2088
22. Ohyashiki JH, Ohyashiki K, Fujimura T, et al (1994) Telomere shortening associated with disease evolution patterns in myelodysplastic syndromes. Cancer Res 54: 3557-3560
23. Ohyashiki JH, Ohyashiki K, Toyama K, et al (1996) A non-radioactive fluorescence-based telomeric repeat amplification protocol to detect and quantitate telomerase activity. Trends Genet 12: 395-396
24. Ohyashiki JH, Iwama H, Yahata N, et al (1999) Telomere stability is frequently impaired in high-risk groups of patients with myelodysplastic syndromes. Clin Cancer Res 5: 1155-1160
25. Li B, Yang J, Andrews C, et al (2000) Telomerase activity in preleukemia and acute myelogenous leukemia. Leuk Lymphoma 36: 579-587
26. Boultwood J, Fidler C, Kusec R, et al (1997) Telomere length in myelodysplastic syndromes. Am J Hematol 56: 266-271
27. Takauchi K, Tashiro S, Ohtaki M, et al (1994) Telomere reduction of specific chromosome translocation in acute myelocytic leukemia. Jpn J Cancer Res 85: 127-130
28. Yamada O, Oshimi K, Motoji T, et al (1995) Telomeric DNA in normal and leukemic blood cells. J Clin Invest 95: 1117-1123
29. Ohyashiki JH, Ohyashiki K, Iwama H, et al (1997) Clinical implications of telomerase activity levels in acute leukemia. Clin Cancer Res 3: 619-625
30. Engelhardt M, Mackebzie K, Drullinsky P, et al (2000) Telomerase activity and telomere length in acute and chronic leukemia, pre- and post-ex vivo culture. Cancer Res 60: 610-617
31. Ball SE, Gibson FM, Rizzo S, et al (1998) Progressive telomere shortening in aplastic anemia. Blood 91: 3582-3592
32. Boultwood J, Peniket A, Watkins F, et al (2000) Telomere length shortening in chronic myelogenous leukemia is associated with reduced time to accelerated phase. Blood 96: 358-361
33. Li J-J, Kook H, Chung I-J, et al (2001) Telomere length changes in patients with aplastic anemia. Br J Haematol 112: 1025-1030
34. Brummendorf TH, Maciejewski JP, Kak J, et al (2001) Telomere length in leukocyte subpopulations of patients with aplastic anemia. Blood 97: 895-900

35. Akiyama M, Hoshi Y, Sakurai H, et al (1998) Changes of telomere length in children after hematopoietic stem cell transplantation. Bone Marrow Transplant 21: 167-171
36. Notaro R, Cimmino A, Tabarini D, et al (1997) In vivo telomere dynamics of human hematopoietic stem cells. Proc Natl Acad Sci USA 94: 13782-13785
37. Wynn RF, Cross MA, Hatton C, et al (1998) Accelerated telomere shortening in young recipients of allogeneic bone-marrow transplants. Lancet 351: 178-181
38. Lee J, Kook H, Chung I, et al (1999) Telomere length changes in patients undergoing hematopoietic stem cell transplantation. Bone Marrow Transplant 24: 411-415
39. Rufer N, Brummendorf TH, Chpuis B, et al (2001) Accelerated telomere shortening in hematological lineages is limited to the first year following stem cell transplantation. Blood 97: 575-577
40. Allsopp RC, Cheshier S, Weissman IL. (2001) Telomere shortening accompanies increased cell cycle activity during serial transplantation of hematopoietic stem cells. J Exp Med 193: 917-924
41. Shay JW, Wright WE (2000) The use of telomerized cells for tissue engineering. Nat Biotechnol 18: 22-23
42. Thomas M, Yang L, Hornsby PJ (2000) Formation of functional tissue from transplanted adrenocortical cells expressing telomerase reverse transcriptase. Nat Biotechnol 18: 39-42
43. Steinert S, Shay JW, Wright WE (2000) Transient expression of human telomerase extends the life span of normal human fibroblasts. Biochem Biophys Res Commun 273:1095-1098

Immunosuppression in Aplastic Anemia

André Tichelli[1] and Gérard Socié[2]

1 Division of Hematology, Kantonsspital, Petersgraben 4, 4031, Basel, Switzerland
2 Service d'Hématologie-Greffe de Moelle, Hôpital Saint Louis, 1 Avenue Claude Vellefaux 75475, Paris CEDEX 10, France

Summary: Immunosuppression is the treatment of choice for patients not eligible for bone marrow transplantation. Currently, the best results are obtained with a combination of ATG and cyclosporine. This combination provides better remission rates, improves disease free survivals, leads to faster improvement in blood counts and is therefore associated with shorter periods of risk from severe cytopenia. The androgens and corticosteroids in normal or high dose as an adjunct to this combination have not further improved the results. About 20-30% of the patients fail to respond to an initial course immunosuppression and further 30% relapse after an initial response to immunosuppression. More than half of them respond to a second course of ATG, without increase in side effects due to this treatment. How patients with aplastic anemia respond to immunosuppressive treatment is independent of their age; however, age does influence survival, with older patients having an increased mortality. Nevertheless, there is no reason to withhold treatment in older patients solely because of their age. The decision to treat older patients should depend on the medical condition of the individual patient rather than a fixed age limit. New approaches are actually under investigation. Cyclophosphamide may induce stable remission. However, it is associated with a prolonged severe cytopenia, with significant more infections and an excess in death rates. G-CSF as an adjunct to ATG and CSA significantly increases neutrophil counts and is associated with fewer days of fever. However, the effect is limited to the time of treatment with growth factors. The results concerning long-term benefit in patients treated with G-CSF are still controversial. Currently, the EBMT is conducting a prospective trial comparing ATG and CSA with or without G-CSF. Thus, immunosuppression has shown to be a real alternative for patients in whom transplantation is not appropriate. However, there are still deaths occurring during the early phase of the treatment, relapses of the disease are common and late complications may be observed. Therefore, new drugs as well as alternative treatment strategies need to be evaluated.

Key words: Aplastic anemia, immunosuppression, antithymocyte globulin, cyclosporine

Introduction

The aim of treatment of patients with aplastic anemia is to improve peripheral blood counts in order that the patients become independent from transfusions and free from risks of opportunistic infections. Immunosuppression is the treatment of choice for patients not eligible for bone marrow transplantation. In contrast to transplantation, where the defective organ is replaced by a healthy marrow, immunosuppressive treatment will not cure the disease but rather aims to eliminate „autoaggressive" cells responsible for the aplastic marrow resulting in pancytopenia.

Today, immunosuppressive regimens for aplastic anemia include antithymocyte globulin (ATG) and cyclosporine. The use of ATG goes back to early studies of Mathé. He used ATG as part of the conditioning regimen before marrow infusion in four accidentally irradiated patients [1]. Although no engraftment was observed, two of the four patients survived presenting an autologous regeneration of hematopoiesis. Similar observations were made by Speck and Kissling in rabbits with benzene-induced aplasia [2]. Control animals died, whereas animals treated with ATG and infusion of HLA-haploidentical marrow recovered autologous bone marrow function. From these observations, it was obvious that ATG alone is not sufficient to allow engraftment of allogeneic marrow. Nevertheless, hemopoietic recovery seemed possible after treatment with ATG. The evidence that autologous reconstitution can occur under experimental and clinical circumstances opened new possibilities in the treatment of patients with aplastic anemia.

Immunosuppressive Regimens

Antithymocyte Globulin (ATG)

Clinical effectiveness of ATG was first apparent in individual patients, where evidence of autologous hemopoietic reconstitution was demonstrated [3]. Since that time, a number of large single-center and multicenter retrospective studies evaluating ATG in the treatment of aplastic anemia have been published in Europe and the USA. The criticism that most of the responses after ATG are spontaneous remissions can no longer be sustained.

The efficacy of ATG was definitively demonstrated by two prospective trials. In the UCLA study, 42 newly diagnosed patients with severe aplastic anemia were randomized to receive ATG or to be followed as controls [4]. Only patients who received ATG responded. At three month, 11 of 21 patients receiving ATG had sustained improvement in hematopoiesis, compared to none of 21 control patients. Twelve control patients subsequently received ATG, and six improved. Overall 53% of the patients treated with ATG responded eventually. The second trial, conducted by the Aplastic Anemia Study Group randomized ATG, HLA-haploidentical marrow, and androgen versus androgen alone [5]. Regimen including ATG resulted in a superior 2-year probability of survival of 76% compared to 31% in the control group. As in the previous trial, response rate was higher in the ATG treated group. Thus, the evidence that the treatment of ATG improves hematopoietic response and survival was definitively established.

Further studies showed a strong correlation between severity of the disease and survival after immunosuppression. Of 64 patients treated with ATG between 1980 and 1985 in the Hammersmith Hospitals, overall actuarial survival at 6 years was 53%, with a 79% survival for nonsevere, and 36% for severe aplastic anemia. The neutrophil and platelet counts before treatment were highly predictive, whereas gender, age, and etiology were not. A retrospective study of the EBMT evaluated survival as a function of disease severity, by correlating disease severity with age for patients treated with immunosuppression or bone marrow transplantation [6]. This study first demonstrated that immunosuppression is a true alternative treatment to bone marrow transplantation. Overall six-year survival among patients of both types of treatment is similar. Furthermore, the analysis showed, that immunosuppression is superior to transplantation for patients with neutrophil counts above 0.2×10^9/L and who are older than 20 years of age. In contrast, outcome of bone marrow transplantation is better for patients with neutrophil counts below $<0.2 \times 10^9$/L and who are younger than .

Cyclosporine

The development of Cyclosporine as a new immunosuppressive drug raised the possibility of its use as an alternative, less toxic treatment of patients with aplastic anemia. Small studies performed during the 1980s suggested that cyclosporine alone was effective in some patients with aplastic anemia who were either untreated or who had failed ATG treatment.

The French Cooperative Study, which compared ATG with cyclosporine as first-line therapy for severe aplastic anemia proved the efficacy of cyclosporine. One hundred nineteen patients were randomized to receive ATG or cyclosporine. At 3-months, patients who had none or minimal response received the alternative therapy to assess the value of a sequential immunosuppression for treatment of severe aplastic anemia. The response rate at 3 months was 11.6% in the cyclosporine arm and 16% in the ATG arm. Survival at 12 months was similar for both types of treatment, 70% in the cyclosporine group, and 64% in the ATG group. From this study it appeared that cyclosporine is as effective as ATG, but that response rates at 3 months were lower than previously reported. The main complication was infection, which occurred more frequently and was more likely to be lethal during ATG therapy [7-9].

Intensification of Immunosuppression

Would intensification of the immunosuppression improve results? This question first was addressed in the German Aplastic Anemia Study Group. Since best results in patients not eligible for bone marrow transplantation are achieved with ATG or cyclosporine, it seemed reasonable to investigate the combination of both immunosuppressants. In a randomized multicenter trial, ATG and methylprednisolone with or without cyclosporine were compared [10]. Eighty four patients were included in the study, 43 in the cyclosporine and 41 in the control arm. At 3 months, the response was 65% for ATG and cyclosporine compared with 39% for ATG, and at 6 months 70% and 46%, respectively. Time to achieve a response was faster after the combination treatment compared to ATG alone, however, the groups did not differ in the quality of remission. The actuarial survival of all patients at 41 months was 64% in the cyclosporine group and 58% in the control arm (p=0.16). The inability to translate the increased response rate into prolonged survival is not really surprising. Indeed non-responders to ATG were crossed over to receive the combination treatment. At last follow-up of this study, patients treated with ATG and cyclosporine showed a clear advantage for failure free survival but presented similar survival probabilities [11].

The efficacy of combined ATG and cyclosporine has also been confirmed for patients with non-severe aplastic anemia and neutrophil counts $\geq 0.5 \times 10^9/L$. In a randomized study, patients treated with cyclosporine alone, administered as out-patient were compared to patients treated with cyclosporine and ATG [12]. This study demonstrated a significantly higher response rate for ATG and cyclosporine (74%) compared to cyclosporine alone (46%) at 6 months, better failure free survival, better blood counts and a lower need for a second course of immunosuppression for patients treated with ATG and cyclosporine. Twenty five percent of patients receiving cyclosporine alone required salvage therapy with ATG because of disease progression before 6 months, while only 6% of patients with ATG and cyclosporine required re-treatment. Despite the different response rates, survival was similar in both types of treatment. The efficacy of the combined therapy, ATG and cyclosporine, has been demonstrated in several other trials[13] and also been proven for children [14,15].

These findings confirm, that ATG is essential in the treatment of aplastic anemia, and that ATG and cyclosporine are the most effective combination. ATG and cyclosporine showed better remission rates, improved disease free survivals as well as faster improvement in blood counts and therefore shorter periods of risk form severe cytopenias. This combination represents therefore the gold standard, to which new drugs or treatment strategies have to be compared in patients, not eligible for bone marrow transplantation.

Adjuncts to ATG and Cyclosporine

Corticosteroids have been used in the treatment of aplastic anemia, long before ATG. No controlled studies have been carried out to investigate whether adding corticosteroids to ATG improve the recovery. Nevertheless, it has been shown in a retrospective analysis, that ATG and corticosteroids were not superior to ATG alone [16]. The use of high-dose methylprednisolone was reported to be effective in some patients with aplastic anemia [17,18]. This led to the general incorporation of high-dose corticosteroids into ATG regimens [19-24]. Later on, high dose methylprednisolone was compared in a controlled randomized study to normal dose of corticosteroids in patients treated with ATG and androgens. Response rates as well as survival probabilities were similar in both treatment arms. There was no difference in acute toxicity, causes of death or late hematological complications [25]. However, at long term, corticosteroids produced more aseptic necrosis of femur [26]. Today, the use of high-dose corticosteroids is no longer justified and normal dose of corticosteroids should given just for a short period to reduce toxicity of ATG infusion and serum sickness [27].

Before ATG became available for the treatment of aplastic anemia, androgens was the unique drug with proven efficacy in some patients with marrow failure. Therefore, despite conflicting results on the potential value of androgens as an adjunct to ATG therapy, most ATG protocols included androgens. In a prospective study from the University of California at Los Angeles the addition of androgens did not improve response rates and survival to ATG [28]. In contrast, a study from Frankfurt showed that patients given ATG combined with androgens responded better, compared with those not given androgens, but had similar survival [29]. A more recent, large prospective randomized study from the EBMT clarified this issue [30]. Patients were given ATG and methylprednisolone for one month, and randomized either to receive (n = 69) or not receive (n = 65) oxymetholone, administered daily for 4 months. The response rate at three months was significantly higher in patients receiving androgens (56% versus 40%), particularly in females with neutrophil counts less than 0.5 x 10^9/L. Survival was comparable in both groups (71% versus 65%). Today, with the improvement of the results by the use of intensive immunosuppressive regimens, and because of long term treatment toxicity of androgens, most protocols do no longer include androgens. However, this drug can be considered for patients failing to respond or for patients where response can not be sustained with cyclosporine.

Cyclophosphamide

Some aplastic anemia patients conditioned with cyclophosphamide for bone marrow transplantation, showed an autologous marrow recovery after graft rejection [31-33]. In addition, recovery from aplastic anemia after treatment with cyclophosphamide alone have been reported [34]. However, cyclophosphamide was considered ineffective in a survey conducted by the American Society of Hematology, revealing only 3 remissions among 73 treated patients from 50 centers [35]. Doses of cyclophosphamide ranged from 100mg daily for up to 4 months to 30mg/kg daily for 4 days. Twenty five percent of patients received cyclophosphamide in doses approaching those used for conditioning prior to marrow transplantation. The three aplastic anemia patients achieving a hematological remission received moderate high-dose of cyclophosphamide (40-120mg/kg) without bone marrow transplantation.

More recently, 10 patients with severe aplastic anemia and who did not have an HLA-identical sibling were treated with high-dose cyclophosphamide, receiving 45mg per kg per day over four consecutive days, with or without cyclosporine [36]. A complete remission was achieved in seven patients. Six patients were alive and in continuous remission 7.3 to 17.8 years after treatment, one has died from the acquired immunodeficiency syndrome 44 months after treatment with cyclophosphamide. The high-dose cyclophosphamide was well tolerated. In responders, a neutrophil count of 0.5×10^9/L was achieved at median 95 days, and the last platelet concentrate was transfused 85 days after treatment. None of the responders relapsed or developed a clonal disorder.

As a result of these findings, a prospective randomized study comparing cyclophosphamide with ATG and cyclosporine was initiated in the USA, but this study has been prematurely closed because of an excess of deaths due to infection during periods of severe neutropenia. At time of interruption of the study, patients treated with cyclophosphamide had significant more days of hospitalization, days of intravenous antibiotics, number of fungal infections, number of platelet and red blood cell transfusions and longer periods of severe neutropenia [37]. Today, cyclophosphamide can not be considered as the first line treatment in newly diagnosed aplastic anemia. Even after relapse or in patients not responding to first course of ATG, it is safer to administer a second or even a third course of ATG, since ATG is a well established treatment modality with recognized but manageable side effects.

The Use of G-CSF

The rationale for the use of growth factors in aplastic anemia is to reduce the risk of infection during the severe neutropenia, to improve the rate and quality of response to immunosuppression, and to induce long term remission of the disease. G-CSF was used either alone or in combination with androgens to treat aplastic anemia. During treatment, neutrophil counts improved significantly in more than 80% of patients, but returned to baseline levels as soon as the growth factors were stopped. In some of these patients bacterial or fungal infections present at the start were resolved during G-CSF treatment.

In a pilot study of the EBMT, 40 patients were treated with G-CSF in combination with ATG and CSA. At 3 months a trilineage response rate of 82% was achieved. The actuarial survival was 92%, with a maximum follow-up of 3 years [38]. An update performed in 2000 confirmed the high probability for patients with aplastic anemia of becoming transfusion independent and surviving after treatment with ALG, CSA, Methylprednisolone, and G-CSF [39]. One important question is, whether G-CSF is required to achieve this excellent survival. The only way to answer this question is with a prospective randomized trial. A trial with a small number of patients has shown earlier neutrophil recovery and fewer infections, but has failed to show a higher rate of trilineage response or survival in the G-CSF-treated group [40]. A second prospective multicenter trial of children with newly diagnosed aplastic anemia was conducted in Japan. All children received ATG, CSA and Danazol. Children with very severe aplastic anemia received G-CSF. All other children were randomized to receive G-CSF (n=35) or not (n=34). Response rate, incidence of febrile episodes and documented infections as well as survival were similar between the different treatment groups, however, the number of patients was small as well [41].

The results concerning long term benefit in patients treated with G-CSF are still not conclusive. There is a concern that chronic use of G-CSF in patients with aplastic anemia may increase the risk of myelodysplastic syndrome and acute leukemia [42-46]. This topic is discussed in detail in the Chapter entitled "Myelodysplastic syndrome, acute leukemia & PNH after immunosuppressive therapy for severe aplastic anemia". At present, there is no indication for the routine use of G-CSF in combination with ATG and cyclosporine outside of multicenter trials. The EBMT is currently conducting a prospective trial comparing ATG and CSA with or without G-CSF including a larger number of patients with acquired aplastic anemia.

Response to Immunosuppression

Timing and Quality of Response

After immunosuppressive treatment response is usually slow, and in many responders blood counts never completely normalize. Platelet and neutrophil counts are lower than normal, red blood cells show a persistent marcrocytosis, there is an increased amount of fetal hemoglobin and bone marrow morphology continues to show stigmata of the disease, such as dysplasia of hemopoiesis, lymphocytosis and increased number of mast cells[10,47,48]. Accordingly, most patients have residual abnormalities of hemopoiesis in vitro, with a reduced growth capacity of hemopoietic progenitors [49,50], as well as a decreased capacity to initiate long-term cultures [51,52].

The time taken to respond to immunosuppression is very variable, ranging form 1 to 12 months. Neutrophil counts generally increase by 2 months after treatment, but patients continue to improve up to 1 year after treatment. Of those who do respond, 45% have done it by three months, and 70% by six months. The pattern of hematological recovery was evaluated in 103 patients with severe aplastic anemia treated with ATG [53]. During and shortly after treatment neutrophil counts rose transiently, probably because of the concomitant use of high-dose corticosteroids, and then fell back to pretreatment values. Thereafter, neutrophil counts increased gradually over nearly 50 weeks, and reticulocyte counts over 8 weeks. Platelet counts were slowest to recover, but an overall increase was observed over nearly 2 years.

Relapse of Aplastic Anemia

Relapse after initial response to immunosuppression have been described soon after the efficacy of ATG therapy has been established [22,54-56]. Today, we know that relapse after immunosuppression is common. Relapse can be defined as a decrease in any of the peripheral blood counts to less than 50% of the median sustained during remission, by a return of the counts to levels fulfilling criteria of severe aplastic anemia or the recurrence of transfusion dependence. A multicenter study conducted by the EBMT SAA Working Party retrospectively evaluated the relapse of aplastic anemia after immunosuppression [57]. Of the 358 responding patients, 74 patients relapsed at mean 25 months after initial immunosuppression. The actuarial risk of relapse was 35% at 14 years. The risk of relapse was higher in patients responding within 120 days from immunosuppression (48%) compared to patients responding between 120 and 360 days (40%) and only 20% for slow responders (>360 days). Age, gender or severity of the disease were not predictive for relapse. The relapses occurred at any time, from few months up to 10 years after immunosuppressive treatment, and there was no particular period of highest occurrence. Thirty nine of the 74 (53%) relapsing patients responded to a second course of immunosuppression. The actuarial survival of relapsing patients, responding to a second treatment is similar to that of patients who respond but do not relapse.

Cyclosporine containing regimens seem to have higher relapse rates than treatments with ATG alone. The relapse rate of the German multicentre trial, at 8.7 years was 40% , including all patients who were treated with cyclosporine at presentation or as salvage. However, the outcome of relapsing patients is not significantly compromised. More than 50% of them respond to a re-treatment with ATG. Some patients are clearly dependent on cyclosporine. In these patients, discontinuation of cyclosporine leads a decrease of neutrophil and platelet counts, and re-institution may induce a secondary response. In the German trial, 29% of the patients who responded to ATG and cyclosporine, required continuous treatment with cyclosporine. Similarly, in a prospective study on 100 patients with aplastic anemia treated with ATG and CSA, with or without G-CSF, the actuarial probability at 5 years of discontinuing CSA is 38%, suggesting CSA dependence in many patients [39]. Relapse and the need for maintenance treatment indicate that patients with aplastic anemia are not cured from their disease after treatment with immunosuppression.

Failure to Respond to Immunosuppression

Despite improvement of response rate with intensive immunosuppression, still 20% to 30% of the patients fail to respond to the initial course of treatment. Today, there is no consensus on treatment for patients not responding to ATG. Since treatment protocols including ATG yield the best results irrespective of additional immunosuppressive drugs, a second trial with ATG appears a logical approach [10,58]. However, doubts about repeating a previously unsuccessful treatment, fear of anaphylactic reactions and serum sickness after repeated exposure to ATG as well as concerns about over-immunosuppression prevented systematic use of repeated courses of ATG for long time [13,59-63]. For second course, it is possible to use either an alternative animal source of ATG or to repeat treatment with an ATG from the same species. The immediate risk of repeated ATG has been evaluated in a single center study [64]. This study showed, that ATG can be repeated in aplastic anemia patients with no higher risks than for initial therapy. Acute reactions of immediate hypersensitivity type are tolerable after re-exposure and reversible in all cases. Incidence and severity of serum sickness is similar in naive and previously exposed patients. Manifestations of serum sickness appear significantly earlier after repeated exposure (median day 6) than after initial exposure. This earlier occurrence of serum sickness after repeated exposures is compatible with a second set response. Due to the retrospective nature of the study, the question, whether sequential courses of ATG might improve response rates and consolidate the remission without increasing the risk of infections or the development of late clonal complications remains unanswered. It is likely that good risk patients were selected for repeated treatment, whereas bad risk patients were not. In a small multicenter prospective trial, one versus two courses of ATG were compared in twenty two patients. Both treatment modalities were well tolerated and both had similar response rates [15]. However, the small number of patients does not allow an interpretation on the efficacy of the treatments. The value of sequential ATG is currently investigated in the prospective, randomized multicenter EBMT trial.

Response According to Age

In allogeneic bone marrow transplantation, age is a major risk factor for transplant related mortality. As a consequence, an upper age limit for transplantation of patients with aplastic anemia is set in many centers between the age of 40 and 50. Immunosuppressive treatment with ATG and cyclosporine is therefore commonly used as first line strategy for older patients. The outcome in patients older than 50 years of age treated with immunosuppression for aplastic anemia has been retrospectively evaluated by the EBMT, Severe Aplastic Anaemia Working Party [65]. The study included 810 patients from 56 centers, treated between 1974 and 1997. Patients were evaluated according to age groups: 127 patients ≥ 60 years, 115 patients between 50 and 59 years, and 568 patients <50years (reference group). Response to immunosuppression, relapse rate, and risk of clonal complication were similar in all three age groups. Age was associated with an increased risk of death due to bleeding or infections, and survival rates were 50% (≥60 years), 75% (50-59 years) and 78% (<50 years), respectively. To evaluated the contribution of expected age-related deaths to the decrease of survival in older patients, the mortality rate was compared with that of a general population matched for sex and age. Compared to a general population, the highest ratios were observed in all age groups during the first year after treatment. Compared with younger patients, older patients had lower standardized mortality ratio. Therefore, to some extend at least, the decrease of survival among older patients is due to the deaths that can be expected in older persons in a general population.

This study shows that immunosuppressive treatment is feasible and effective in patients of 50 years of age or older. There is no reason to withhold treatment in older patients with aplastic anemia solely because of their age. The decision to tread older patients should depend more on the medical condition of the individual patient rather than a fixed age limit.

Lessons from immunosuppressive Treatment for PNH Patients

It is now well recognized that patients with aplastic anemia often have a coexisting PNH clone. A PNH clone may occur in 15-50% of aplastic anemia patients [66-68]. From most observations of aplastic anemia patients treated with immunosuppression, the presence of such a clone seems not to affect the chance of response and the survival. Studies with a small number of patients described a response to the treatment with cyclosporine alone or with ATG [69-71]. G-CSF may induce the expression of Fc gamma R-III (CD16) in aplastic anemia/PNH patients with neutrophils affected by paroxysmal nocturnal hemoglobinuria [72]. In a single center study, the correlation of flowcytometric data were correlated with diverse marrow failure syndromes. Evidence of PNH was found in 25 of 115 patients with aplastic anemia, and in 9 of 39 (23%) patients with myelodysplastic anemia. The response rate in aplastic anemia patients with or without PNH clone was similar, 71% and 82% respectively. The presence of a PNH clone in myelodysplastic syndrome predicted responsiveness to immunosuppressive treatment [73]. Thus, immunosuppression is effective to correct pancytopenia due to marrow failure in PNH patients. However, there is no evidence that the risk of thrombosis or hemolytic crises are modified by this treatment modality.

Immunosuppression has shown to be a real alternative for patients with bone marrow failure due to aplastic anemia or paroxysmal nocturnal hemoglobinuria in whom transplantation is not appropriate. However, there are still deaths occurring during the early phase of the treatment, relapses of the disease are common and late complications may be observed. Therefore, new drugs as wells as alternative treatment strategies need to be evaluated.

References

1. Mathé G, Amiel JL, Schwarzenberg L, Choay J, Trolard P, Schneider M, Hayat M, Schlumberger JR, Jasmin CL (1970) Bone marrow graft in man after conditioning with antilymphocyte serum. British Medical Journal. 2:131-136
2. Speck B, Kissling M (1971) Successful bone marrow grafts in experimental aplastic anaemia using antilymphocyte serum for conditioning. Eur J Clin Biol Res. 10:1047-1051
3. Jeannet M, Speck B, Rubinstein A, Pelet B , Wyss M, Kummer H (1976) Autologous marrow reconstitutions in severe aplastic anaemia after ALG pretreatment and HLA-A semi-incompatible bone marrow cell transfusion. Acta Haematologica. 55:129-139
4. Champlin RE, Ho W, Gale RP (1983) Antilymphocyte globulin treatment in patients with aplastic anemia. A prospective randomized trial. N Engl J Med. 308:113-118
5. Camitta B, O'Reilly JO, Sensenbrenner LL, Rappeport JM, Champlin RE, Doney K, August C, Hoffmann R, Kirkpatrick D, Stuard R, Santos G, Parkman R, Gale RP, Nathan DG (1983) Antithoracic duct lymphocyte globulin therapy of severe aplastic anemia. Blood. 62:883-888
6. Bacigalupo A, Hows J, Gluckman E, Nissen C, Marsh JCW, Van Lint MT, Congiu M, De Planque MM, Ernst P, McCann S, Ragavashar A, Frickhofen N, Würsch A, Marmont AM, Gordon-Smith EC (1988) Bone marrow transplantation (BMT) versus immunosuppression for the treatment of severe aplastic anaemia: A report of the EBMT SAA Working Party. Br J Haematol. 70:177-182
7. Gluckman E, Esperou-Bourdeau H, Baruchel A, Boogaerts M, Briere J, Donadio D, Leverger G, Leporrier M, Reiffers J, Janvier M (1992) Multicenter randomized study comparing cyclosporine-A alone and antithymocyte globulin with prednisone for treatment of severe aplastic anemia. Blood. 79:2540-2546
8. Gluckman E, Esperou-Bourdeau H, Baruchel A, Boogaerts M, Briere J, Donadio D, Leverger G, Leporrier M, Reiffers J, Janvier M (1992) A multicenter randomized study comparing cyclosporin-A alone and antithymocyte globulin with prednisone for treatment of severe aplastic anemia. The cooperative group on the treatment of aplastic anemia. J Autoimmun. 5 Suppl A:271-5:271-275
9. Esperou H, Devergie A, Lehn P, Lallemand A, Gluckman E (1989) A randomized study comparing ciclosporin A and antithymocyte globulin for treatment of severe aplastic anemia. Nouv Rev Fr Hematol. 31:65-68
10. Frickhofen N, Kaltwasser JP, Schrezenmeier H, Ragavashar A, Vogt HG, Herrmann F, Freund M, Meusers P, Salama A, Heimpel H (1991) Treatment of aplastic anemia with antilymphocyte globulin and methylprednisolone with or without cyclosporine. N Engl J Med. 324:1297- 1304
11. Frickhofen N, Rosenfeld SJ (2000) Immunosuppressive treatment of aplastic anemia with antithymocyte globulin asnd clyclosporine. Seminars in Hematology. 37:56-68
12. Marsh J, Schrezenmeier A, Marin P, Ilhan O, Ljungman P, McCann S, Socie G, Tichelli A, Passweg J, Hows J, Raghavachar A, Locasciulli A, Bacigalupo A (1999) Prospective randomized multicenter study comparing cyclosporin alone versus the

combination of antithymocyte globulin and cyclosporin for the treatment pf patients with nonsevere aplastic anemia: a report from the European Blood and Marrow Transplant (EBMT). Blood. 93:2191-2195
13. Rosenfeld SJ, Kimball J, Vining D, Young NS (1995) Intensive immunosuppression with antithymocyte globulin and cyclosporine as treatment for severe acquired aplastic anemia. Blood. 85:3058-3065
14. Führer M, Burdach S, Ebell W (1998) Relapse and clonal disease in children with aplastic anemia after immunosuppressive therapy: the SAA 94 experience. German/Austrian Pediatric Aplastic Anemia Working Group. Klin Padiatr. 210:173-179
15. Matloub YH, Smith C, Bostrom B, Koerper MA, O'Leary M, Khuder S, Smithson WA, Nickerson HJ, Silberman T, Hilden J, Moertel CL, Month S, Monteleone P, Ramsay NK (1997) One course versus two courses of antithymocyte globulin for the treatment of severe aplastic anemia in children. J Pediatr Hematol Oncol. 19:110-114
16. Young N, Speck B (1984) Antithymocyte and antilymphocyte globulins: clinical trials and mechanism of action. Prog Clin Biol Res. 148:221-6:221-226
17. Bacigalupo A, Giordano D, Vimercati R, Marmont AM (1979) Bolus methylprednisolone in severe aplastic anemia. N Engl J Med. 300:501-502
18. Bacigalupo A, Van Lint MT, Cerri R, Giordano D, Santini G, Carrella M , Damasio EE, Rossi E, Risso M, Vimercati R, Podesta M, Durando A, Reali G, Avanzi G, Barbanti M, Marmont AM (1980) Treatment of severe aplastic anemia with bolus of 6-methylprednisolone and antilymphocyte globulin. Blut. 41:168-171
19. Novitzky N, Wood L, Jacobs P (1991) Treatment of aplastic anaemia with antilymphocyte globulin and high- dose methylprednisolone. Am J Hematol. 36:227-234
20. Novitzky N, Mobara G, Jacobs P (1992) Antilymphocyte globulin and high-dose methylprednisolone improve survival in patients with aplastic anaemia without additional financial costs. S Afr Med J. 81:254-257
21. Kojima S, Miyajima Y, Fukuda M, Matsuyama T, Maeda H, Yamamoto K, Tsuzuki S, Akatsuka M, Sugihara T, Minami S (1992) [Treatment of aplastic anemia with antilymphocyte globulin, high-dose methylprednisolone and androgen]. Rinsho Ketsueki. 33:11-16
22. Hinterberger-Fischer M, Hinterberger W, Hockern P, Schmidmeier W, Gadner H, Geissler K, Kos M, Schwarzinger I, Neumann E, Niessner H (1986) Treatment of severe aplastic anemia with combined immunosuppression (antithymocyte globulin and high-dose methylprednisolone). Acta Haematol. 76:196-201
23. Doney K, Storb R, Buckner CD, McGuffin R, Witherspoon RP, Deeg HJ, Appelbaum F, Sullivan KM, Thomas ED (1987) Treatment of aplastic anemia with antithymocyte globulin, high-dose corticosteroids, and androgens . Exp Hematol. 15:239-242
24. Speck B, Gratwohl A, Nissen C, Osterwalder B, Würsch A, Tichelli A, Lori A, Reusser P, Jeannet M, Signer E (1986) Treatment of severe aplastic anemia. Exp Hematol. 14:126-132
25. Doney K, Pepe M, Storb R, Bryant E, Anasetti C, Appelbaum FR, Buckner CD, Sanders J, Singer J, Sullivan KM, Weiden P, Hansen J (1992) Immunosuppressive treatment of aplastic anemia: Results of a prospective randomized trial of antilymphocyte globulin (ATG), methylprednisolone, and oxymetholone to ATG, very high-dose methylprednisolone, and oxymetholone. Blood. 79:2566-2571

26. Doney K, Leisenring W, Storb R, Appelbaum FR (1997) Primary treatment of acquired aplastic anemia: Outcomes with bone marrow transplantation and immunosuppressive therapy. Ann Intern Med. 126:107-115
27. Bacigalupo A, Van Lint MT, Congiu M, Pittaluga PA, Occhini D, Marmont AM (1986) Treatment of SAA in Europe 1970-1985: A report of the SAA Working Party. Bone Marrow Transplant. 1:611-615
28. Champlin RE, Ho WG, Feig SA, Winston DJ, Lenarsky C, Gale RP (1985) Do androgens enhance the response to antithymocyte globulin in patients with aplastic anemia? A prospective randomized trial. Blood. 66:184-188
29. Kaltwasser JP, Dix U, Schalk KP, Vogt H (1988) Effect of androgens on the response to antithymocyte globulin in patients with aplastic anaemia. Eur J Haematol. 40:111-118
30. Bacigalupo A, Chaple A, Hows J, Van Lint MT, McCann S, Milligan D, Chessells J, Goldstone AH, Ottolander J, Van't Veer ET, Comotti B, Coser P, Broccia G, Bosi A, Locascuilli A, Catalano L, Battista R, Arcese W, Carotenuto M, Marmont AM, Gordon-Smith EC (1993) Treatment of aplastic anaemia (AA) with antilymphocyte globulin (ALG) and methylprednisolone (MPred) with or without androgens: A randomized trial from the EBMT SAA working party. Br J Haematol. 83:145-151
31. Speck B, Cornu P, Jeannet M, Nissen C, Burri HP, Groff P, Nagel GA, Buckner CD (1976) Autologous marrow recovery following allogeneic marrow transplantation in a patient with severe aplastic anemia. Exp Hematol. 4:131-137
32. Sensenbrenner LL, Steele AA, Santos G (1977) Recovery of hematologic competence without engraftment following attempted bone marrow transplantation for aplastic anemia. Exp Hematol. 5:51-
33. Territo MC (1977) Autologous bone marrow repopulation following high dose cyclophosphamide and allogeneic marrow transplantation in aplastic anemia. Br J Haematol. 36:305-
34. Baran DT, Griner PF, Klemperer MR (1976) Recovery from aplastic anemia after treatment with cyclophosphamide. N Engl J Med. 295:1522-
35. Griner PF (1980) A survey of the effectiveness of cyclophosphamide in patients with severe aplastic anemia. Am J Hematol. 8:55-60
36. Brodsky RA, Sensenbrenner LL, Jones RJ (1996) Complete remission in severe aplastic anemia after high-dose cyclophosphamide without bone marrow transplantation. Blood. 87:491-494
37. Tisdale JF, Dunn DE, Geller N, Plante M, Nunez O, Dunbar CE, Barrett AJ, Walsh TJ, Rosenfeld SJ, Young NS (2000) High-dose cyclophosphamide in severe aplastic anaemia: a randomised trial. Lancet. 356:1554-1559
38. Bacigalupo A, Broccia G, Corda G, Arcese W, Carotenuto M, Gallamini A , Locatelli F, Mori PG, Saracco P, Todeschini G (1995) Antilymphocyte globulin, cyclosporin, and granulocyte colony- stimulating factor in patients with acquired severe aplastic anemia (SAA): a pilot study of the EBMT SAA Working Party. Blood. 85:1348-1353
39. Bacigalupo A, Bruno B, Saracco P, Di Bona E, Locasciulli A, Locatelli F, Gabbas A, Dufour C, Arcese W, Testi G, Broccia G, Carotenuto M, Coser P, Barbui T, Leoni P, Ferster A (2000) Antilymphocyte globulin, cyclosporine, prednisolone, and granulocyte colony-stimulating factor for severe aplastic anemia: an update of the GITMO/EBMT study on 100 patients. European Group for Blood and Marrow Transplantation (EBMT) Working Party on Severe Aplastic Anemia and the Gruppo Italiano Trapianti di Midolio Osseo (GITMO). Blood. 95:1931-1934

40. Gluckman E, Rokicka-Milewska R, Gordon-Smith E, et al (1998) Results of a randomized study of glycosylated rHuG-CSF lenograstim in severe aplastic anemia (abstract). 92:1551-
41. Kojima S, Hibi S, Kosaka Y, Yamamoto M, Tsuchida M, Mugishima H, Sugita K, Yabe H, Ohara A, Tsukimoto I (2000) Immunosuppressive therapy using antithymocyte globulin, cyclosporine, and danazol with or without human granulocyte colony-stimulating factor in children with acquired aplastic anemia. Blood. 96:2049-2054
42. Hashino S, Imamura M, Tanaka J, Kobayashi S, Musashi M, Kasai M, Asaka M (1996) Transformation of severe aplastic anemia into acute myeloblastic leukemia with monosomy 7. Ann Hematol. 72:337-339
43. Kojima S, Tsuchida M, Matsuyama T (1992) Myelodysplasia and leukemia after treatment of aplastic anemia with G-CSF. N Engl J Med. 326:1294-
44. Ohara A, Kojima S, Hamajima N, Tsuchida M, Imashuku S, Ohta S, Sasaki H, Okamura J, Sugita K, Kigasawa H, Kiriyama Y, Akatsuka J, Tsukimoto I (1997) Myelodysplastic syndrome and acute myelogenous leukemia as a late clonal complication in children with acquired aplastic anemia. Blood. 90:1009-1013
45. Ohsaka A, Sugahara Y, Imai Y, Kikuchi M (1995) Evolution of severe aplastic anemia to myelodysplasia with monosomy 7 following granulocyte colony-stimulating factor, erythropoietin and high-dose methylprednisolone combination therapy [see comments]. Intern Med. 34:892-895
46. Yamazaki E, Kanamori H, Taguchi J, Harano H, Mohri H, Okubo T (1997) The evidence of clonal evolution with monosomy 7 in aplastic anemia following granulocyte colony-stimulating factor using the polymerase chain reaction. Blood Cells Mol Dis. 23:213-218
47. Tichelli A, Gratwohl A, Nissen C, Signer E, Stebler Gysi Ch, Speck B (1992) Morphology in patients with severe aplastic anemia treated with antilymphocyte globulin. Blood. 80:337-345
48. Marsh JCW, Hows J, Bryett KA, Al-Hashimi S, Fairhead SM, Gordon-Smith EC (1987) Survival after antilymphocyte globulin therapy for aplastic anemia depends on disease severity. Blood. 70:1046-1052
49. Nissen C, Cornu P, Gratwohl A, Speck B (1980) Peripheral blood cells from patients with aplastic anaemia in partial remission suppress growth of their own bone marrow precursors in culture. Br J Haematol. 45:233-243
50. Rudivic R, Jovcic G, Bilijanovic-Paunovic L, Stojanovic N, Mijoric A, Partivic Kentera V (1985) Myelopoiesis and erythropoiesis of bone marrow cells cultured in vitro in patients recovered from aplastic anaemia. Br J Haematol. 70:55-62
51. Marsh JC, Chang J, Testa NG, Hows J, Dexter TM (1990) The hematopoietic defect in aplastic anemia assessed by long-term marrow culture. Blood. 76:1- 10
52. Alexanian R, Nadell J, Alfrey C (1972) Oxymetholone treatment for anemia of bone marrow failure. Blood. 40:353-365
53. Nissen C, Gratwohl A, Tichelli A, Stebler C, Wursch A, Moser Y, dalle Carbonare V, Signer E, Buser M, Ritz R (1993) Gender and response to antilymphocyte globulin (ALG) for severe aplastic anaemia. Br J Haematol. 83:319-325
54. Fairhead SM, Chipping PM, Gordon-Smith E (1983) Treatment of aplastic anaemia with antilymphocyte globulin (ALG). Br J Haematol. 55:7-16
55. Speck B, Gratwohl A, Nissen C, Leibundgut U, Ruggero O, Osterwalder B, Burri HP, Cornu P, Jeannet M (1981) Treatment of severe aplastic anaemia with

antilymphocyte globuline or bone marrow transplantation. British Medical Journal. 282:860-863
56. Miller WJ, Branda RF, Flynn PJ, Howe RB, Ramsay NKC, Condie RM, Jacob HS (1983) Antilymphocyte globulin treatment of severe aplastic anaemia. Br J Haematol. 55:17-25
57. Schrezenmeier H, Marin P, Raghavachar A, McCann S, Hows J, Gluckman E, Nissen C, van't Veer-Korthof ET, Ljungman P, Hinterberger W, Van Lint MT, Frickhofen N, Bacigalupo A, for the EBMT SAA Working Party (1993) Relapse of aplastic anaemia after immunosuppressive treatment: a report from the European Bone Marrow Transplantation Group SAA Working Party. Br J Haematol. 85:371-377
58. Young NS, Barrett AJ (1995) The treatment of severe acquired aplastic anemia. Blood. 85:3367-3377
59. Doney K, Kopecky K, Storb R, Buckner CD, Singer J, Anasetti C, Appelbaum F, Beatty P, Bensinger W, Berenson R, Clift R, Deeg HJ, Hansen J, Hill R, Loughran T, Martin P, Petersen FB, Sale G, Sanders J, Sullivan KM, Stewart P, Weiden P, Witherspoon RP, Thomas ED (1990) Long-term comparison of immunosuppressive therapy with antithymocyte globulin to bone marrow transplantation in aplastic anemia. In: Shahidi NT (eds) Aplastic anemia and other bone marrow failure syndromes. Springer-Verlag Inc, New York Berlin Heidelberg London Paris Tokyo, pp 104-114
60. Means RT, Jr., Krantz SB, Dessypris EN, Lukens JN, Niblack GD, Greer JP, Flexner JM, Stein RS (1988) Re-treatment of aplastic anemia with antithymocyte globulin or antilymphocyte serum. Am J Med. 84:678-682
61. Marsh JC, Socie G, Schrezenmeier H, Tichelli A, Gluckman E, Ljungman P, McCann SR, Raghavachar A, Marin P, Hows JM (1994) Haemopoietic growth factors in aplastic anaemia: a cautionary note. European Bone Marrow Transplant Working Party for Severe Aplastic Anaemia [see comments]. Lancet. 344:172-173
62. Marsh JCW, Gordon-Smith EC (1988) The role of antilymphocyte globulin in the treatment of chronic acquired bone marrow failure. Blood Rev. 2:141-148
63. Socie G, Henry-Amar M, Bacigalupo A, Hows J, Tichelli A, Ljungman P, McCann SR, Frickhofen N, Van't Veer-Korthof E, Gluckman E (1993) Malignant tumors occurring after treatment of aplastic anemia. European Bone Marrow Transplantation-Severe Aplastic Anaemia Working Party. N Engl J Med. 329:1152-1157
64. Tichelli A, Passweg JR, Nissen C, Bargetzi M, Hoffmann T, Wodnar-Filipowicz A, Signer E, Speck B, Gratwohl A (1998) Repeated treatment with horse antilymphocyte globulin for severe aplastic anaemia. Br J Haematol. 100:393-400
65. Tichelli A, Socie G, Henry-Amar M, Marsh J, Passweg J, Schrezenmeier H, McCann S, Hows J, Ljungman P, Marin P, Raghavachar A, Locasciulli A, Gratwohl A, Bacigalupo A (1999) Effectiveness of immunosuppressive therapy in older patients with aplastic anemia. European Group for Blood and Marrow Transplantation Severe Aplastic Anaemia Working Party. Ann Intern Med. 130:193-201
66. Schrezenmeier H, Hertenstein B, Wagner B, Raghavachar A, Heimpel H (1995) A pathogenetic link between aplastic anemia and paroxysmal nocturnal hemoglobinuria is suggested by a high frequency of aplastic anemia patients with a deficiency of phosphatidylinositol glycan anchored proteins [published erratum appears in Exp Hematol 1995 Feb;23(2):181]. Exp Hematol. 23:81-87

67. Tichelli A, Gratwohl A, Würsch A, Nissen C, Speck B (1988) Late haematological complications in severe aplastic anaemia. Br J Haematol. 69:413-418
68. Najean Y, Haguenauer O (1990) Long-term (5 to 20 years) Evolution of nongrafted aplastic anemias. The Cooperative Group for the Study of Aplastic and Refractory Anemias. Blood. 76:2222-2228
69. Schubert J, Scholz C, Geissler RG, Ganser A, Schmidt RE (1997) G-CSF and cyclosporin induce an increase of normal cells in hypoplastic paroxysmal nocturnal hemoglobinuria. Ann Hematol. 74:225-230
70. Jego P, Le Strat A, Girard L, Sebillot M , Grosbois B, Le Blay R, Drenou B (1997) Paroxysmal nocturnal hemoglobinuria: efficacy of prolonged treatment with granulocyte colony-stimulating factor. Blood. 90:2841-2843
71. van Kamp H, van Imhoff GW, de Wolf JT, Smit JW, Halie MR, Vellenga E (1995) The effect of cyclosporine on haematological parameters in patients with paroxysmal nocturnal haemoglobinuria. Br J Haematol. 89:79-82
72. Ninomiya H, Muraki Y, Shibuya K, Nagasawa T, Abe T (1993) Induction of Fc gamma R-III (CD16) expression on neutrophils affected by paroxysmal nocturnal haemoglobinuria by administration of granulocyte colony-stimulating factor. Br J Haematol. 84:497-503
73. Dunn DE, Tanawattanacharoen P, Boccuni P, Nagakura S, Green SW, Kirby MR , Kumar MS, Rosenfeld S, Young NS (1999) Paroxysmal nocturnal hemoglobinuria cells in patients with bone marrow failure syndromes. Ann Intern Med. 131:401-408

Part 7

PNH, Aplastic Anemia, Myelodysplastic Syndromes (2): Clinical Aspects

Myelodysplastic syndrome, acute leukemia & PNH after immunosuppressive therapy for severe aplastic anemia

Gérard Socié, MD, Ph.D.[1] and André Tichelli[2]

[1] Service d'Hématologie-Greffe de Moelle, Hôpital Saint Louis, 1 Avenue Claude Vellefaux 75475, Paris CEDEX 10, France
[2] Départment of Hematology, Kantonsspital, Petersgraben 4, 4031, Basel, Switzerland

Summary: Recent progress in the treatment of aplastic anemia (AA) has dramatically changed the previously grim prognosis for these patients. Improvements in immunosuppressive therapy have increase the number of long-term survivors so that immediate survival is no longer the sole concern. Here, we review the major clinical studies and summarize recent analyses of risk factors for developing paroxysmal nocturnal hemoglobinuria (PNH), myelodyplastic syndrome (MDS), and acute leukemia after treatment for aplastic anemia. We also review biological clues that may shed some light on the interrelationship between AA and clonal diseases.

Key words: Aplastic anemia, long-term, malignancies, clonal diseases

Introduction

Treatment of aplastic anemia has dramatically changed the previously grim prognosis for these patients, so that today the 2-year survival is 60–90%. Improved results following immunosuppression increase the number patients alive years after presentation so that immediate survival is no longer the sole concern. Indeed, quality of life and late complications are becoming increasingly important.

After immunosuppression, recovery often takes several months and the stigmata of the disease persist despite clinical remission. Although immunosuppression allows autologous reconstitution of hemopoiesis, it leaves the patient with a fragile bone marrow prone to the development of a clonal disorder, such MDS, acute leukemia, or PNH.

MDS and acute leukemia

Early clinical studies

In 1988 two single-center studies described the evolution of acquired aplastic anemia into MDS or leukemia in patients treated with immunosuppression. In Leiden (1), five of 35 aplastic anemia patients surviving for two or more years developed a MDS. All five had had a normal karyotype at the time of first diagnosis. Overall, 10% of all patients surviving for 2 years eventually developed a leukemic disorder. Similarly, the Basel group reviewed late hematological disorders in 137 patients with aplastic anemia (2). One hundred and three patients were treated with ALG. Twenty of the 103 patients treated with immunosuppression showed a clonal evolution (8 MDS; 12 PNH; 1 MDS together with PNH). The risk of developing a hematological complication after immunosuppressive treatment was 57% at 8 years. In a subsequent follow-up study, it was found that cytogenetic analysis at diagnosis was normal in 13 patients who developed MDS/leukemia (3).

Long-term follow-up was evaluated in a first retrospective multicenter European Bone Marrow Transplantation (EBMT) group study including 468 aplastic anemia patients treated with ALG (4). Of the 245 patients who survived for 2 years, 21 later died. Between the second and third years, pancytopenia due to aplastic anemia was still the major cause of death of the long-term survivors; after the third year posttreatment, leukemia and MDS became the primary cause of death. Twelve patients developed MDS/leukemia at a median of 4.6 years after immunosuppression. The actuarial mortality of long-term survivors was 22% at 8 years, but so far no plateau has occurred. The 5-year actuarial probability of developing MDS was between 15% and 20%.

To examine the incidence of hematological cancers in a large cohort of long-term survivors after treatment of aplastic anemia, the EBMT Severe Aplastic Anemia (SAA) Working Party studied the cases reported to the registry. The study population consisted of 860 patients treated with immunosuppression (5). The risk of malignancy was analyzed overall and according to treatment relative to the risk in the general population. Thirty-four of the 860 patients who received immunosuppressive treatment developed MDS or leukemia (19 MDS, 15 leukemia). The risk factors for MDS or acute leukemia after immunosuppressive treatment included older age (relative risk or RR = 1.03), being treatment in or after 1982, as compared with in or before 1981 (RR = 3.01), having a splenectomy (RR = 3.65), and multiple courses of immunosuppression (RR = 2.26). The addition of androgens to the immunosuppressive treatment was associated with a lower risk of malignancy (RR = 0.28). This study definitively proved the increased risk of developing MDS and leukemia after immunosuppressive treatment.

MDS and acute leukemia in the era of intensive immunosuppressive therapy

MDS & acute leukemia after ALG and cyclosporine

The above mentioned studies included mainly patients who were treated with ALG alone. However, standard treatment of SAA nowadays includes the association of ALG with cyclosporine (CSA) [for review, see (6) and the chapter by A.Tichelli, this volume]. Data on MDS and leukemia in paptients who received ALG plus CSA are still few.

Unpublished updated results from the NIH (Rosenfeld and Young, personnal communication April 99) on 121 patients who received ATG and cyclosporine as first line therapy are of special interest since this combination is now the standard therapy in patients with severe aplastic anemia. G-CSF was not a routine part of the treatment regimen. Nine patients have evolved to myelodyspalstic syndrome (8 with clonal chromosomal abnormalities). The risk for such evolution was 7% at 2 years and 15% at 5 years. Long-term analysis of the German study group trial comparing ALG versus ALG plus CSA found 4 cases of MDS/leukemia within the entire group of 84 patients, with no apparent increase of malignancies in patients who received the intensive arm (6). Finally, the UCLA group reported a 13% projected incidence of hematological malignancies (5/155 cases) in a cohort of patients who were treated with ALG $\pm$ Androgens $\pm$ CSA(7).

Does G-CSF increase the risk of hematological malignancies after immunosuppressive therapy ?

Some recent findings have suggested that administration of G-CSF was associated with the development of MDS/leukemia. These data come mainly from Japan (8). In a study of 167 children with aplastic anemia, 11 of 50 patients treated with G-CSF and cyclosporine developed MDS/leukemia (9). Of the group of children treated with G-CSF or cyclosporine alone, none had this complication. The 7-year probability of developing MDS/leukemia was 47±17% for patients treated with growth factors as compared to 16±6.% for the whole cohort. Furthermore, patients treated for longer periods or with a higher cumulative growth factor dose seemed to be at an increased risk. Eight of the 11 late clonal disorders developed within 36 months, much earlier than is usually reported (8).

Recently, however two papers provided less alarming estimates of secondary hematological malignancies in SAA patients who received G-CSF as supportive therapy. The Italian group reported data on a large group of patients with aplastic anemia (n=144) (10); Eligible patients had severe aplastic anemia and were receiving ALG and CSA, with or without the addition of human recombinant G-CSF. All patients had a normal baseline cytogenetic analysis either before or immediately after treatment with antilymphocyte globulin, and at least one subsequent cytogenetic analysis at follow-up; 57 patients who were treated

Table 1. Incidence of MDS / AML after intensive immunosuppressive therapy for aplastic anemia

Trial /study	Reference	Treatment	N° of MDS-AL/ Total	Projected incidence
UCLA	Paquette, 1995	ATG/MP ±A ± CSA	5 / 155	13%
JCAASG	Ohara, 1997	Various, 18 with G.CSF	11 / 119	47%
Tokyo	Kaito, 1998	Various, 18 with G.CSF	5 / 72	Not available
German group	Frickhofen,	ATG+MP / ATG+MP+CSA	4 / 84	Not available
NIH	Rosenfeld,	ATG+MP+CSA	9 / 121	15%
JCAASG	Kojima, 2000	ATG/MP +A + CSA ± G.CSF	13 / 144	9%
IAASG	Locasciulli, 2001	ATG/MP + CSA ± G.CSF	3 / 119	3%

UCLA; university California / Los Angeles: JCAASG; Japanese children aplastic anemia study group: NIH; Nationale Institute of Health, USA: IAASG; Italian aplastic anemia study group.
ATG; anti-thymocyte globulins: MP; methyl prednisolone: A; androgens: CSA; cyclosporine: MDS; myelodysplastic syndrome: AML; acute myeloid leukemia.

between 1978 and 1991and had never been given growth factors (non-G-CSF group) had a median follow-up of 3157 days. 87 patients who were treated between 1991 and 1996 and received G-CSF in combination with immunosuppressive treatment (G-CSF group), had a follow-up of 1390 days. The median duration of G-CSF treatment was 6·5 months (3-48) and the median total dose received was 36 000 g/patient (1200–141 000 g). A malignant disorder was diagnosed during follow-up in 13 (9%) of 144 patients: eight developed MDS (3 in G-CSF and 5 in non-G-CSF group), 4 had acute leukemia (two in each group). The interval between immunosuppressive treatment and occurrence of malignant complications was 380 days for the non-G-CSF group and 765 days (180–1610) days for the G-CSF group (p=0·4). The risk of developing secondary malignancies at 60 months was 9% and 7%, respectively, for patients with or without G-CSF treatment (p=0·99). A prospective Japanese multicenter trial of 119 childrenl to 18 years of age with newly diagnosed AA was conducted, comparing treatment using ATG, CSA, and danazol with or without rhG-CSF (11). All children with very severe AA received rhG-CSF (VSAA group, n=50). The other children were randomized to receive ATG, CyA, DAN, and rhG-CSF (n= 35) or ATG, CyA, and DAN without rhG-CSF (n=34). MDS/AML developed in one patient in each of the three groups; the overall risk for MDS/AML was 3% at 4 years. Data on MDS/AML following intensive immunosuppressive therapy are summarized in **Table 1**.

Cytogenetic abnormalities and aplastic anemia

Cytogenetic analysis can demonstrate clonal changes before overt MDS or leukemia develops. In an early survey from Seatlle, an abnormal karyotype was detected in seven of 183 (4%) patients with otherwise morphologically typical aplastic anemia, three of the seven patients were treated with immunosuppression and two developed late MDS or leukemia (12). Among the 169 patients with typical aplastic anemia without cytogenetic anomalies, five (3%) subsequently presented a leukemic transformation. In an Italian study (13), bone marrow cells from 69 patients with acquired aplastic anemia were analyzed cytogenetically at diagnosis and after immunosuppressive treatment (13): fifty-one patients (74%) were normal and remained so, but 18 patients (26%) had at least one abnormal karyotype. Remarkably, three patterns of evolution were observed. In seven patients the anomaly was acquired after treatment, in three patients it was present at diagnosis and persisted during evolution, and in eight patients it represented a transient post-treatment abnormality. Most frequent abnormalities were trisomy 8 (n=8) and monosomy 7 (n=2). The three patients with persistent cytogenetic abnormalities all developed leukemia.

The EBMT SAA Working Party retrospectively evaluated cytogenetic anomalies in aplastic anemia patients treated with immunosuppression between 1974 and 1994 [A. Tichelli et al. unpublished data]. One hundred and seventy evaluable karyotypes from six European centers were reported for analyses. The median number of metaphases evaluated per karyoptype was 12 (range 6–100). One hundred and forty-six patients had a normal karyotype and one patient had a somatic chromosomal defect [46,XY,t(4;12)(q34;21)], confirmed by family karyotyping. In 23 cases (12%) chromosomal aberration in hemopoietic cells could be demonstrated. The anomaly was clonal in 19 of the 23 (11%) cases. The chromosomal defect concerned all metaphases in six cases; a mosaic of normal and abnormal chromosomes was found in the 17 remaining patient samples. The defect was simple in 18 patients and complex in the remaining one. Trisomy 8 was observed in five cases and wass therefore the most frequent single anomaly. Trisomy 6 was found in two patients, 5q– in another two, and an anomaly on chromosome 7 in three others (monosomy 7; deletion 7; isochromosome 7). All other abnormalities were seen in single cases. Follow-up of the karyotype for a selected group of patients was reported. At diagnosis, four of 31 cases presented a clonal anomaly (one case with monosomy 8); at the first analysis after therapy eight of 23 patients had a clonal anomaly (three cases with monosomy 8), as did 12 of 78 cases at the last analysis (four cases with monosomy 8). Other studies including data on cytogenetics in SAA are summarized in **Table 2**.

In the above mentioned Italian study on 144 patients with SAA (10), follow-up analyses showed abnormal cytogenetics in 13 (23%) of 57 cases in the non-G-CSF group and in 9 (10%) of 87 patients in the G-CSF group. The median interval between immunosuppressive treatment and appearance of cytogenetic abnormalities was shorter in the G-CSF group (740 days) than the non-G-CSF

Table 2. Recent cytogenetic studies in aplastic anemia

Reference	N° of patients with abnormal caryotypes / N available	Most frequent abnormalities
Geary, 1999 (32)	12 / ?	-7, n=1; +8, n=3
Yamazaki, 1997 (33)	11 / ?	-7, n=9
Piaggio, 1999 (34)	13 / 97	+8, n=4

group (1825 days; p=0·03). The risk of developing cytogenetic abnormalities at 5 years was 14% and 11% for the G-CSF and non-GCSF group, respectively (p=0·99). Abnormal cytogenetics were transient in 9 cases (4 in G-CSF, 5 in non-G-CSF) and persistent in 13 (5 in G-CSF, 8 in non-G-CSF, p=0·1).

All these reports show that a single clonal chromosomal defect in hemopoietic cells is not infrequent in aplastic anemia at presentation. Trisomy 8 and monosomy 7 are the most usual aberrations. However, cytogenetic analysis of empty marrow may be technically difficult, and classic karyotyping upon diagnosis of severe aplastic anemia is often not possible because of a lack of metaphases. The interphase fluorescent in situ hybridization (FISH) technique may detect anomalies such as trisomies and monosomies, even in marrow from patients in which no metaphases are found.

Molecular analyses

Clonality studies

Recently, aplastic anemia patients were reported to have molecular evidence of clonal hematopoiesis. But these data must now be interprted in the light of recent clonality studies in normal individuals. Early clonality studies using X-chromosome inactivation methods were published in 1991. In Leiden, 13/18 patients with aplastic anemia were shown to have a unilateral X-inactivation pattern compatible with monoclonality (14). By contrast, only one of 6 patients with AA had such a profile in the St. Georges Hospital study (15). In theses two early studies, no somatic DNA controls were included to exclude a constitutionally skewed pattern of gene methylation. More recent studies raised questions as to the interpretation of such analyses (16;17). No correlation was found between clonality and flow cytometry results in patients with a GPI-molecule defect. In two patients with molecularly sequenced *N-ras* point mutations, one showed a monoclonal pattern but the other one was polyclonal by X-chromosome inactivation techniques. Furthermore, in recent clonality studies in normal individuals nearly 20% of normal women exhibited an X-chromosome inactivation pattern compatible with monoclonality. In addition, and there is often tissue specificity of X-chromosome inactivation pattern and, the proportion of normal women with a monoclonal pattern increased with age (18).

A recent technique developed by the St Georges group allowed analysis of 2 genes polymorphisms (G6PD and IDS genes) by RT-PCR (19). This technic allows to study clonality without problems inherent to DNA methylation. A clonal pattern was found in 10/26 patients with AA (38%), significantly more frequent than found in aged-mached individuals.

Telomere loss in aplastic anemia

Recent reports of progressive telomere shortening in aplastic anemia deserves special comment. In a seminal study, Ball and coworkers studied mean telomere length (TRF) in blood leukocytes from 79 patients with alastic anemia. TRF lengths were significantly shorter for age in patients as compared to normal individuals (20). Telomere shortening was apparent both in granulocyte and mononuclear cells, suggesting loss at the level of the hematopoietic stem cell. There was a significant correlation between telomere loss and disease duration, in addition to age-related loss. Finally, of the 5 patients with TRF less than 5.0 kb, 3 had acquired cytogenetic abnormalities suggesting that telomere erosion may be relvant to the pathogenesis of MDS in aplastic anemia. Similar results have since been confirmed by the NIH group (21), using a FLOW-FISH technic, found that, on separated leukocytes, telomere shortening occurred predominantly in myeloid cells while it was non significantly affected in lymphocyte. Lee and coworkers in another study found that mean TRF was especially shortened in patients non responding to treatment (22).

N-ras gene mutations in patients with aplastic anemia

An exhaustive study by Mortazavi and coworkers looked for *N-ras* gene mutation in a 30 apalstic patients on the basis of evidence for clonal evolution (9 females with monoclonal pattern, 3 with MDS, and 18 with PNH). Only 1 patient with AA/PNH transformed into MDS showed a *N-ras* codon 13 mutation. Of interest, *N-ras* was demonstated to occur at the time of evolution to MDS and developed in non-PNH cells (23).

MDS/AML in aplastic anemia: A model of multistep carcinogenesis (24)

There is growing evidence that hematopoiesis is not only quantitatively but also qualitatively abnormal in aplastic anemia. Abnormal hematopoiesis seems to be clonal in some cases or at least subject to a premature process of cell aging (TRF length shorteneing). This clone has a growth advantage in the aplastic marrow. Then accumulation of mutational events through *N-ras, fms, p53* or other non-identified gene, mutations, and abnormal clonal cytogenetic abnormalities may lead to the progression of the mutated clone and clinical emergence of MDS and

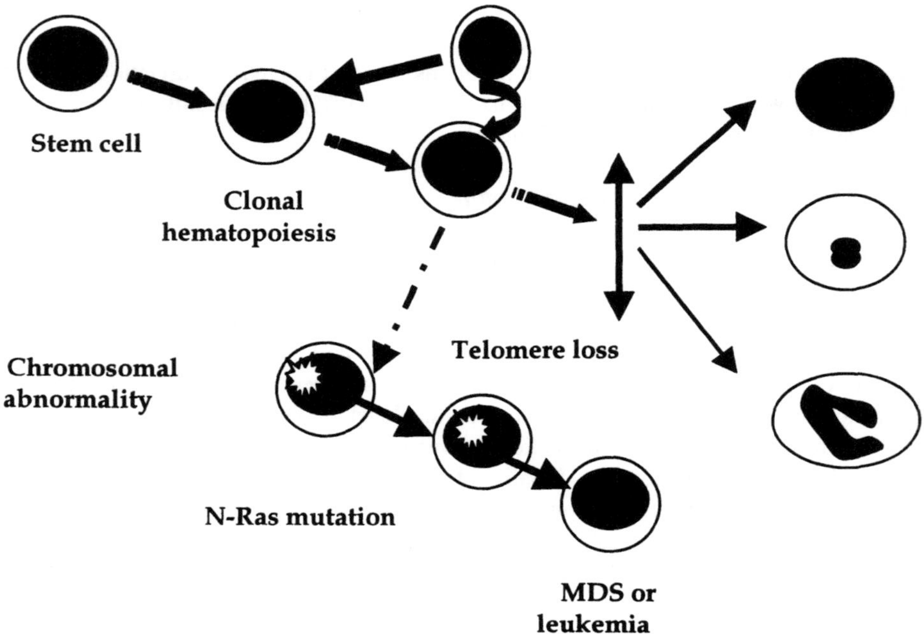

Figure 1. MDS/AML in aplastic anemia: A model of multistep carcinogenesis

then acute leukemia. In this model AA could be compared to polyposis, the first step of colo-rectal carcinogenesis (**Figure 1**).

Is PNH a late clonal disorder ?

Clinical data

In a study of late hematological disorders, including 137 patients with aplastic anemia, 12 of the 103 patients treated with anti-lymphocyte globulin (ALG) developed PNH, and in one patient MDS developed together with PNH (2). Four of the 13 patients with PNH only had biological anomalies, i.e., a positive sucrose test and/or Ham test, while nine patients showed clinical symptoms of the disease. The cumulative risk of developing PNH at 15 years was 25%. At its onset, the disease was usually limited to an abnormal laboratory test, but most patients eventually became symptomatic. The development of PNH has also been evaluated in the large retrospective multicenter EBMT study on the long-term follow-up of 468 aplastic anemia patients treated with ALG. PNH developed in 19 patients, and was diagnosed earlier in the course of the disease, at a median of 3 years after immunosuppression, compared with MDS (at a median of 4.6 years)

(4). The 5-year actuarial probability of developing PNH was between 15% and 20%.

Flow cytometry data

According to immunophenotype analysis, the number of aplastic anemia patients with a clonal PNH defect is much higher than previously suspected. In an early study, GPI-linked protein deficiency was identified in 27 out of 52 (52%) patients with aplastic anemia, affecting at least one cell population (25). Granulocytes were involved in 25, monocytes in 18, lymphocytes in seven, and erythrocytes in seven of the 27 patients with a PNH clone. Normal cells and abnormal cells with a GPI-linked protein deficiency usually coexist, with a mean of 33% cells being affected. Meanwhile, it has become evident that a PNH clone can be present when aplastic anemia is diagnosed; however, these patients do not usually show clinical hemolysis. Rather, flow cytometric analysis of granulocytes and monocytes allows the GPI-linked protein deficiency to be detected (26).

In patients from St. George's Hospital, London, a GPI-anchored protein defect was identified in 17 of 111 (15%) patients with aplastic anemia who had a negative Ham test, 12 patients received bone marrow transplantation, 94 patients were treated with immunosuppression, and 12 patients had neither transplantation nor immunosuppression. GPI-anchored protein expression status prior to immunosuppression did not appear to influence the response rate to treatment (50% response rate in those with GPI-anchored protein deficiency versus 75% in those without). The survival at 12 years after diagnosis was similar in both groups, about 90% (27). A recent study by the NIH group also found a high incidence of GPI-anchored protein defect in patients with aplastic anemia before therapy (5/33, 15%) and most responded to immunosuppressive therapy (28).

Molecular data

Since the description of PIG-A gene alteration in PNH patients, over one hundred PNH cases have been studied at the molecular level [**review in this issue by Takeda** and(29)]. The majority of the genetic alterations are deletions of a few bases resulting in frameshift abnormalities and usually ending with a premature stop codon. A large number of mutations have been located in exon 2 (the largest exon). But more important, on a practical ground, there is no evidence for recurring genetic alterations in PNH, i.e. there is no hot spot for mutations in the PIG-A gene. It is thus highly unlikely that PNH will be amenable to a routine molecular diagnosis. In patients with AA/PNH there is no evidence of different mutations as compared to patients with *de novo* PNH (30;31).

Should PNH still be considered as a late complication ?

In conclusion, it is now clear that a proportion (roughly one third) of patients that would have been considered to have idiopathic aplastic anemia only a few years ago are demonstrated now to have a GPI-anchored protein defect even in the absence of a positive Ham test. Some patients will develop overt PNH in the years following immunosuppressive therapy. At the beginning of the 21st century it may be appropriate to reconsider the problem of PNH and aplastic anemia. In both *de novo* PNH and in the aplastic anemia-PNH syndrome, the PIG-A gene is affected. Both should be regarded as a single disease entity that can be first expressed clinically either as an aplastic anemia or a hemolytic anemia. Thus, even describing PNH as a "late" clonal complication of aplastic anemia may be old-fashioned.

References

1. De Planque MM, Kluinnelemans HC, Van Krieken HJ, et al (1989) Evolution of acquired severe apastic anaemia to myelodysplastic and subsequent leukemia in adults. Br J Haematol **70**:55-61
2. Tichelli A, Gratwohl A, Wursch A, Nissen C, Speck B (1988) Late haematological complications in severe aplastic anaemia. Br J Haematol **69**:413-418.
3. Tichelli A, Gratwohl A, Nissen C, Speck B (1994) Late Clonal Complications in Severe Aplastic Anemia. Leuk Lymphoma **12**:167-175.
4. De Planque MM, Bacigalupo A, Wursch A, et al (1989) Long-term follow-up of severe aplastic anaemia patients treated with antilymphocyte globulin. Br J Haematol **73**:121-126.
5. Socie G, Henryamar M, Bacigalupo A, et al (1993) Malignant Tumors Occurring After Treatment of Aplastic Anemia. N Engl J Med **329**:1152-1157.
6. Socie G, Rosenfeld S, Frickhofen N, Gluckman E, Tichelli A. Late clonal diseases of treated aplastic anemia (2000) Semin Hematol **37**:91-101.
7. Paquette RL, Tebyani N, Frane M, et al (1995) Long-term outcome of aplastic anemia in adults treated with antithymocyte globulin: Comparison with bone marrow transplantation. Blood **85**:283-290.
8. Kaito K, Kobayashi M, Katayama T, et al (1998) Long-term administration of G-CSF for aplastic anaemia is closely related to the early evolution of monosomy 7 MDS in adults. Br J Haematol **103**:297-303.
9. Ohara A, Kojima S, Hamajima N, et al (1997) Myelodysplastic syndrome and acute myelogenous leukemia as a late clonal complication in children with acquired aplastic anemia. Blood **90**:1009-1013.
10. Locasciulli A, Arcese W, Locatelli F, Di Bona E, Bacigalupo A (2001) Treatment of aplastic anaemia with granulocyte-colony stimulating factor and risk of malignancy. Italian Aplastic Anaemia Study Group. Lancet **357**:43-44.
11. Kojima S, Hibi S, Kosaka Y, et al (2000) Immunosuppressive therapy using antithymocyte globulin, cyclosporine, and danazol with or without human granulocyte

colony-stimulating factor in children with acquired aplastic anemia. Blood **96**:2049-2054.

12. Appelbaum F, Barrall J, Storb R, et al (1987) Clonal cytogenetic abnormalities in patients with otherwise typical aplastic anemia. Exp Hematol **15**:1134-1139.
13. Mikhailova N, Sessarego M, Fugazza G, et al (1996) Cytogenetic abnormalities in patients with severe aplastic anemia. Haematologica **81**:418-422.
14. Van Kamp H, Landegent JE, Jansen RPM, lWillemze R, Fibbe WE (1991) Clonal hematopoiesis in patients with acquired aplastic anemia. Blood **78**:3209-3214.
15. Josten KM, Tooze J, Borthwick-Clarke C, Gordon-Smith EC, Rutherford TR (1991) Acquired aplastic anemia and paroxysmal nocturnal hemoglobinuria: studies on clonality. Blood **78**:3162-3167.
16. Raghavachar A, Janssen JWG, Schrezenmeier H, et al (1995) Clonal hematopoiesis as defined by polymorphic X-linked loci occurs infrequently in aplastic anemia. Blood **86**:2938-2947.
17. Cachia PG, Taylor C, Thompson PW, et al (1994) Non-Dysplastic Myelodysplasia. Leukemia **8**:677-681.
18. Gale RE, Fielding AK, Harrison CN, Linch DC (1997) Acquired skewing of X-chromosome inactivation patterns in myeloid cells of the elderly suggests stochastic clonal loss with age. Br J Haematol **98**:512-519.
19. Mortazavi Y, Chopra R, Gordonsmith EC, Rutherford TR (2000) Clonal patterns of X-chromosome inactivation in female patients with aplastic anaemia studied using a novel reverse transcription polymerase chain reaction method. Eur J Haematol **64**:385-395.
20. Ball SE, Gibson FM, Rizzo S, Tooze JA, Marsh JCW, Gordonsmith EC (1998) Progressive telomere shortening in aplastic anemia. Blood **91**:3582-3592.
21. Brummendorf TH, Maciejewski JP, Mak J, Young NS, Lansdorp PM (2001) Telomere length in leukocyte subpopulations of patients with aplastic anemia. Blood **97**:895-900.
22. Lee JJ, Kook H, Chung IJ, et al (2001) Telomere length changes in patients with aplastic anaemia. Br J Haematol **112**:1025-1030.
23. Mortazavi Y, Tooze JA, Gordonsmith EC, Rutherford TR (2000) N-RAS gene mutation in patients with aplastic anemia and aplastic anemia/paroxysmal nocturnal hemoglobinuria during evolution to clonal disease. Blood **95**:646-650.
24. Socie G (1996) Could aplastic anemia be considered a pre-preleukemic disorder? Eur J Haematol **57**:60-63.
25. Schrezenmeier H, Hertenstein B, Wagner B, Raghavachar A, Heimpel H. (1995) A pathogenetic link between aplastic anemia and paroxysmal nocturnal hemoglobinuria is suggested by a high frequency of aplastic anemia patients with a deficiency of phosphatidylinositol glycan anchored proteins. Exp Hematol **23**:181-186.
26. Griscellibennaceur A, Gluckman E, Scrobohaci ML, et al (1995) Aplastic anemia and paroxysmal nocturnal hemoglobinuria: Search for a pathogenetic link. Blood **85**:1354-1363.
27. Delord C, Tooze JA, Saso R, Marsh JCW, Gordonsmith EC (1998) Deficiency of glycosylphosphatidyl inositol-anchored proteins in patients with aplastic anaemia does not affect response to immunosuppressive therapy. Br J Haematol **101**:90-93.
28. Dunn DE, Tanawattanacharoen P, Boccuni P, et al (1999) Paroxysmal nocturnal hemoglobinuria cells in patients with bone marrow failure syndromes. Ann Intern Med **131**:401-408.

29. Socie G.(1997) Recent advances in paroxysmal nocturnal hemoglobinuria. From the biology to the clinic. Hematol Cell Ther **39**:175-187.
30. Nagarajan S, Brodsky RA, Young NS, Medof ME (1995) Genetic defects underlying paroxysmal nocturnal hemoglobinuria that arises out of aplastic anemia. Blood **86**:4656-4661.
31. Pramoonjago P, Pakdeesuwan K, Siripanyaphinyo U, et al (1999) Genotypic, immunophenotypic and clinical features of Thai patients with paroxysmal nocturnal haemoglobinuria. Brit J Haematol **105**:497-504.
32. Geary CG, Harrison CJ, Philpott NJ, Hows JM, Marsh JCW (1999) Abnormal cytogenetic clones in patients with aplastic anaemia: response to immunosuppressive therapy. Brit J Haematol **104**:271-274.
33. Yamazaki E, Kanamori H, Taguchi J, Harano H, Mohri H, Okubo T (1997) The evidence of clonal evolution with monosomy 7 in aplastic anemia following granulocyte colony-stimulating factor using the polymerase chain reaction. Blood Cells Molecules Dis **23**:213-218.
34. Piaggio G, Podesta M, Pitto A, et al (1999) Coexistence of normal and clonal haemopoiesis in aplastic anaemia patients treated with immunosuppressive therapy. Brit J Haematol **107**:505-511.

HEMOPOIETIC STEM CELL TRANSPLANTATION (HSCT) FOR PAROXYSMAL NOCTURNAL HEMOGLOBINURIA (PNH) and ACQUIRED SEVERE APLASTIC ANEMIA (SAA):a report from the SAA Working Party of the European Blood and Marrow Transplant group (EBMT)

Andrea Bacigalupo, Rosi Oneto, Barbara Bruno and Hubert Schrezenmeier on behalf of the working party on severe aplastic anemia (WPSAA) of the European Blood and Marrow Transplantation group (EBMT)

Dipartimento di Ematologia e Oncologia, Ospedale San Martino, Genova, Italy

Abstract: *Background and Objective.* Severe aplastic anemia (SAA) and paroxysmal nocturnal haemoglobinuria (PNH) share many clinical and possibly pathogenetic features Allogeneic hemopoietic stem cell transplantatin (HSCT) is a curative treatment for both disorders. In the present study we compare the outcome of PNH and SAA patients allografted in Europe in the past 3 decades.

Patients. The patient population consisted of 2678 patients with severe aplastic anemia (SAA) and of 90 patients with PNH and reported to the EBMT. The donor was an HLA identical sibling in 2117 SAA and 74 PNH patients respectively, an identical twin in 45 and 0, a mismatched family or matched unrelated donor in 514 and 16 patients. The median time from diagnosis to transplant 92 days for SAA and 882 for PNH ($p<0.0001$) and the median age of patients 18 and 29 respectvely ($p<0.0001$). Radiation was used in the conditioning regimen in 19% and 21% of patients ($p=0.5$); the source of stem cells was bone marrow in 92% and 86% respectively ($p=0.04$).

Results. The overall actuarial survival for HLA identical sibling HSCT 10 years is 62% in SAA vs 58% in PNH ($p=0.09$). Patients with PNH had more acute ($p<0.002$) and more chronic GvHD ($p=0.04$) as compared to SAA patients, also when adjusted for age and excluding radiation based conditioning. There is a strong overall effect of patient age, interval diagnosis-HSCT and year of HSCT on outcome: older age is a significant negative predictor in both SAA and PNH, whereas year of HSCT and duration of the disease appears to be significant only in SAA.

The small number of PNH patients undergoing an alternative donor transplant (n=16) did relatively well with 12 patients surviving. *Conclusion.* This study confirms the feasibility of HSCT in patients with PNH. When compared to the SAA patients, there are differences in patient characteristics, disease duration and outcome of the transplant. Patients receiving a non radiation based conditioning and bone marrow as the source of stem cells, seem to be doing better

INTRODUCTION

A close association between PNH and Aplastic Anemia (AA) is well know: patients with AA often show laboratory evidence of PNH. The proportion of AA with PNH clones varies from 13% to 57% in different studies [1-4] Approximately 15% of previous PNH shows a loss of hematopoietic progenitor cells and then marrow hypoplasia.[3] Stem cells from patients PNH are separated into GPI – AP- and GPI – AP+ populations. The rate of expansion of a PNH clone might be influenced by the existence of an additional marrow injury: PIG-A mutation could be necessary but not sufficient for development of PNH[1] and the relationship between PNH and AA might be explained by a "conditional growth or survival advantage – of PNH cells – in an environment which is injurious to hematopoietic stem cells through a GPI – mediated mechanism."[5]

The clinical course of PNH is highly variable, usually chronic[6]. At diagnosis, a high number of patients (90%) has peripheral blood abnormalities. Thrombotic complications, progression to pancytopenia, myelodysplastic syndrome, and acute leukaemia occur respectively in 30%, 15%, 5%, and 1% of PNH patients within 8 years from diagnosis. Other complications are often treatment-related (haemochromatosis, hepatitis).[7]

The survival at 10 years, reported by several authors, ranges from 50% to 71%.[8-10] The prognostic factors for survival are: thrombosis as complication, development of pancytopenia, development of MDS or acute leukaemia, age over 55 years, need for additional treatment, thrombocytopenia at diagnosis.[7] Thrombosis, infections and haemorrhage are the primary causes of death of PNH patients.[8-10] There are different therapeutic approaches for PNH: corticosteroids, androgens, red blood transfusions, immunosuppressive drugs (antilymphocyte globulin, cyclosporin), and anticoagulation which does not cure the disease, but could decrease the complications[11-15]. Allogeneic hemopoietic stem cell transplantation (HSCT) may cure PNH[16-21], but is associated with transplant related toxicity and mortality.

We report here the EBMT experience of 90 PNH patients who underwent an allogeneic HSCT in Europe between 1976 and 1999, compared with a concurrent larger group of patients with severe aplastic anemia (SAA).

METHODS

Patients were reported to the the Working Party on Severe Aplastic Anemia (WPSAA) of the European Blood and Marrow Transplantation group (EBMT) ,

Table 1. Clinical data of patients allografted for SAA or PNH

	SAA	PNH
Number of transplants	2957	97
Number patients	2678	90
HLA identical siblings	2117	74
Identical twins	45	0
Unrelated donors	425	15
Family HLA phenotyp.id.	91	1

Abbreviations: SAA=acquired severe aplastic anemia; PNH= paroxysmal nocturnal hemoglobinuria

and registered prospectively since 1979. Rosi Oneto has been in charge of the Registry for the past 15 years.

Patients

Clinical data of patients are outlined in Table 1. There were 3054 transplants of which 2642 1st transplants. Of these the large majority were from HLA identical siblings. PNH was the primary diagnosis in 90 patients, whereas acquired SAA was the primary diagnosis in 2678 patients.

Transplants

The donor was an HLA identical sibling in 2117 SAA and 74 PNH patients respectively, an identical twin in 45 and 0, a mismatched family or matched unrelated donor in 514 and 16 patients. The median number of cells infused was 3.2x10^8/kg (0.8-17) in SAA and 3.4x10^8/kg (0.25-12) in PNH . The total number of transplants was 3054 given to a total of 2768 patients. We will consider only first transplants in this report.

Statistics

The Chi square test was used for 2x2 tables, and actuarial survival plots for survival analysis: the end point was survival. A COX multivariate analysis was run with the following factors: patient age, patient gender, interval diagnosis-HSCT, year of transplant; the end point was death due to any cause.

Table 2. Clinical data of patients receiving a 1st HSCT from HLA identical siblings

	SAA	PNH	
Number of patients	2117	74	
Recipient median age in years	18 (1-69)	28 (3-46)	0.0001
Gender male/female	1291/807	43/31	0.6
Donor median age in years	19 (1-71)	28 (2-52) 0	.0001
Gender male/female	1105/921	32/39	0.4
Interval Diagnosis-HSCT	97	882	
	(29-7987)	(32-8676)	0.0001
Conditioning regimen with			
Radiation (TBI/TAI/TLI)	497 (19%)	19 (21%)	0.5
Median n.cells infused x10^8/kg	3.2(0.8-17)	3.4(0.25-12)	0.6
Median follow up in days	1556	1263	
For surviving patients (range)	(156-9006)	(20-4879)	0.02

Abbreviations: TBI= total body irradiation; TLI= total lymphoid irradiation; TAI= thoraco abdominal irradiation

RESULTS and DISCUSSION

Engraftment

Engraftment occurred in the SAA patients in 84%, 88%, 91% respectively in the years <=1990, <=1995, <=1999 respectively , with rejection being recorded in 10%, 9%, 7% (p=0.001). For patients with PNH these figures were 77%, 84%, 88% and rejection in 16%,16% 12% in the years <=1990, <=1995, <=1999 respectively , and rejection in 10%, 9%, 7% (p=0.001). There was no difference in the engraftment rate for patients with SAA and PNH. It should be noted that lack of engraftment is being reported in approximately 7% of the patients also in recent transplant era.

Graft versus host disease (GvHD)

The risk of developing acute GvHD grade 0-I, II, III-IV was 71,14,16% for SAA and 59, 25, 16% for PNH (p=0.05). To rule out that the difference may be due to difference in age and conditioning regimen we analysed only patients under the age of 30, grafted beyond 1992 and with no radiation: the proportion of patients with GvHD grade 0-I, II, III-IV was 82,12,16% for SAA and 50, 22, 27% for PNH (p=0.0002). There was also a borderline increase in chronic GvHD (p=0.04). The reasons for the increase in GvHD are not clear: this may be due the longer duration of the disease or to the older patient age, which may trigger GvHD. And as expected both acute (p<0.0001) and chronic (p<0.0001) GvHD had a profound and significant effect on survival. It is consequential that patients with more acute and/or chronic GvHD should have an excess risk of death.

Survival

The overall actuarial 10 year survival for HLA identical sibling HSCT is 62% in SAA vs 54% in PNH (p=0.09). In a COX multivariate analysis predictive factors were found to be patient age, interval diagnosis-HSCT and year of HSCT . When looking at the effect of these 3 factors, age was a significant predictor separately in SAA and PNH. On the contrary the predictive effect interval diagnosis-HSCT and especially of the year of transplant was evident only in SAA patients. Therefore the difference in outcome between SAA and PNH have become apparent in recent rather than earlier years.

Conditioning regimen

Patients receiving a radiation free conditioning had an advantage over patient receiving radiation (75% vs 65%, p=0.1) and the same could be said for patients receiving BM (77%) vs PB grafts (62%)(p=0.08). We then looked at patients receiving BM grafts without irradiation and survival was 75% vs 60% for patients receiving PB and radiation (p=0.04). So best results are still obtained with the use of BM and no radiation and this should probably be the basis for a prospective transplant trial in PNH.

Alternative donor transplants

The small number of PNH patients undergoing an alternative donor transplant (n=16) did relatively well with 12 patients surviving, proving the procedure is feasible, though the median follow up of surviving patients is short (442 days).

CONCLUSIONS

In this registry based analysis, PNH patients were significantly older than SAA patients and were grafted with a significantly longer interval from diagnosis. There seems to be an excess risk of death following allogeneic HSCT in patients with PNH when compared to SAA patients. This is not due to worse engraftment, but to more severe acute and chronic GvHD. There has been no improvement of outcome in the past 2 decades in PNH in contrast with the very significant improvement of results in acquired SAA. Also this lack of improvement is poorly understood. Reduced intensity conditioning regimens may improve the outcome in the near future.

The report confirms that allogeneic HSCT can produce long term disease free survival in patients with PNH, also in the haemolytic type of the disease, and should be considered before the occurrence of disease or transfusion related complications.

Although most patients have been prepared with the combination of cyclophosphamide and either radiation or an alkylating agent such as busulfan, the use of cyclophosphamide alone is a possibility which should be tested. Indeed cyclophopshamide alone may be sufficient to allow engraftment of allogeneic HLA identical cells: this is especially true after the demonstration that clonality in PNH patients does not necessarily translate in malignancy, and that normal individuals do have granulocytes with the point mutation at the PIG-A gene level.[23]

Acknowledgments. This work was supported by BIOMED EEC grant 951031 to HS, Associazione Italiana Ricerca contro il Cancro (A.I.R.C.) Milano grant to AB and Associazione Ricerca Trapianto Midollo Osseo (A.RI.T.M.O.) Genova. We are grateful to all European Centers who have contributed their data.

REFERENCES

1. Parker CJ. Molecular basis of Paroxysmal Nocturnal Hemoglobinuria. Stem Cells 1996;**14**: 396–411.
2. Rosse WF Ware RE. The molecular basis of paroxysmal nocturnal haemoglobinuria. Blood 1995; **86**: 3277-86
3. Schrezenmeier H, Hertenstein B, Wagner B, et al. A pathogenetic link between aplastic anemia and paroxysmal nocturnal hemoglobinuria is suggested by a high frequency of aplastic anemia patients with a deficiency of phosphatidylinositol glycan anchored proteins. Exp Hematol 1995; **23**: 81-7.
4. Griscelli-Bennaucer A, Gluckman E, Scrobohaci ML, et al. Aplastic Anemia and Paroxysmal Nocturnal Hemoglobinuria: Search for a Pathogenetic Link. Blood 1995; **85**:1354-63.
5. Luzzatto L, Bessler M, Rotoli B. Somatic mutations in Paroxysmal Nocturnal Hemoglobinuria: a Blessing in Disguise?. Cell 1997; **88**: 1-4.
6. Spath-Schwalbe E, Schrezenmeier H, Heimpel SH. Paroxysmal nocturnal hemoglobinuria. Clinical experiences with 40 patients at one center over 25 years. Dtsch Med Wochenschr 1995;**28**: 1027-33.
7. Socié G, Mary J, de Gramont A, et al. Paroxysmal nocturnal haemoglobinuria: long-term follow-up and prognostic factors. Lancet 1996;**348**: 573-8.
8. Dacie JV, Lewis M. Paroxysmal nocturnal hemoglobinuria: clinical manifestation, haematology, and nature of the disease. Series Haematol 1972; **3**: 3-23.
9. Hillmen P, Lewis SM, Bessler M, Luzzato L, et al. Natural history of paroxysmal nocturnal hemoglobinuria. N Engl J Med 1995; **9**: 1253-8.
10. Fujioka S, Takayoshi T. Prognostic features of paroxysmal nocturnal hemoglobinuria in Japan. Acta Haematol JPN 1989; **52**: 1386-94.

11. Packman CH. Pathogenesis and management of paroxysmal nocturnal haemoglobinuria. Blood Rev 1998; 12(1): 1-11.
12. Van Kamp H, Van Imhoff, de Wolf JT, et al. The effect of cyclosporine on haematological parameters in patients with paxysmal nocturnal haemoglobinuria. Br J Haematol 1995; **89** (1): 79-82.
13. Stoppa AM, Vey N, Sainty D, et al. Correction of aplastic anaemia complicating paroxysmal nocturnal haemoglobinuria: absence of eradication of the PNH clone and dependence of response on cyclosporin A administration. Br J Haematol 1996; **93** (1): 42-4.
14. Schubert J, Scholz C, Geissler RG, et al. G-CSF and cyclosporin induce an increase of normal cells in hypoplastic paroxysmal nocturnal hemoglobinuria. Ann Hematol 1997; **74**: 225-30.
15. Paquette RL, Yoshimura R, Veiseh C, et al. Clinical characteristics predict response to anthithymocyte globulin in paroxysmal nocturnal haemoglobinuria. Br J Haematol 1997; **96**: 92-7.
16. Szer J, Deeg HJ, Witherspoon RP, et al. Long-term survival after marrow transplantation for paroxysmal nocturnal hemoglobinuria with aplastic anemia. Ann Intern med 1984; **101**: 193-5.
17. Antin JH, Ginsburg D, Smith BR, et al. Bone marrow transplantation for paroxysmal nocturnal hemoglobinuria: eradication of the PNH clone and documentation of complete lymphohematopoietic engraftment. Blood 1985; **66**: 1247-50.
18. Kolb HJ, Holler E, Bender-Gotze, Walther U. Myeloablative conditioning for marrow transplantation in myelodysplastic syndromes and paroxysmal nocturnal haemoglobinuria. Bone Marrow Transplant 1989; **4**: 29-34.
19. Kawahara K, Witherspoon SP, Storb R. Marrow transplantation for paroxysmal nocturnal hemoglobinuria. Am J Hematol 1992; **39**: 283-8.
20. Bemba M, Guardiola P, Garderet et al. Bone marrow transplantation for paroxysmal nocturnal haemoglobinuria. Br J Haematol 1999, **105** : 366 – 8.
21. Saso R, Marsh J, Cevreska L et al. Bone marrow transplants for paroxysmal nocturnal haemoglobinuria. . Br J Haematol 1999, **104** : 392 – 6.
22. Locasciulli A. Hepatitis viruses infections in bone marrow transplantation. Bone Marrow Transplant 1996; **18** Suppl 2: 115 - 6.
23. Araten DJ, Nafa K, Pakdeesuwan K, Luzzatto L. Clonal populations of hematopoietic cells with paroxysmal nocturnal hemoglobinuria genotype and phenotype are present in normal individuals. Proc Natl Acad Sci USA 1999, **27**: 5209 – 14

Part 8

Related Topics

Mutation Analysis of *PIG-A* Gene in Korean Patients with Paroxysmal Nocturnal Hemoglobinuria

J H Yoon[1], H I Cho[1], S S Park[1], Y H Chang[1], B K Kim[2]

[1]Department of Clinical Pathology, [2] Department of Internal Medicine, Seoul National University Hospital, 28 Yongun-dong, Chongno-gu, Seoul 110-744, Korea

Summary: The characteristics of somatic mutation of the PIG-A gene in Korean patients with PNH were studied. Twenty four patients with PNH were selected. Ham tests and sucrose hemolysis tests were done for them all. CD59 expressions of erythrocytes for fourteen and granulocytes for five patients were performed to confirm diagnosis. Dideoxy fingerprinting (ddF) was used to screen mutations, and direct sequencing of DNA was performed to characterize the mutations.

Gene mutation was detected in 12 out of the 24 patients. The other 12 patients were negative in ddF screening. Ten novel mutations and two known mutations were detected. Their mutations consisted of five deletions, six substitutions and one insertion. These mutations resulted in six premature terminations, three abnormal splicings, one missense in exon 2 and two nonsense mutations. Two patients with venous thrombosis showed mutations in exon 3 only. The patterns of mutations were substitution mutation in six patients and frameshift mutation in the other six.

There were ten novel mutations among the twelve mutations in the Korean patients with PNH and the characteristics of the mutations varied without any significant hot spot in sites or types.

Key words: Korean, Paroxysmal nocturnal hemoglobinuria(PNH), Mutation, Dideoxy fingerprinting (ddF)

Introduction

PNH is the second most common hemolytic anemia in Korean after autoimmune hemolytic anemia. [1] PNH has been diagnosed on the basis of clinical pictures including intravascular hemolysis with recurrent episodes of emission of dark urine in the morning and following confirmation by Ham test. The molecular test for mutation analysis can also serve as a confirmation test of PNH. Hence this study was performed in order to obtain information on the mutation sites of Korean patients with PNH.

Methods

Twenty four Korean patients with PNH who had been diagnosed based on the Ham test and CD59 expression.

Mutations were screened by dideoxy finger printing (ddF) method, a modified PCR-SSCP. PCR product was purified using the BM High Pure PCR Purification Kit (Roche, Basel, Switzerland) and ddF reaction was performed using the ThermoSequenase Radio Labeled Terminator Cycle Sequencing Kit (Amersham Pharmacia Biotech, Inc., Cleveland, USA). Electrophoresis was done using 0.35mm, 0.5× non-denaturing mutation detection enhancement(MDE) gel.

DNA sequencing was accomplished for the patients screened by ddF to confirm the mutation sites. Direct sequencing was achieved using the ThermoSequenase Radio Labeled Terminator Cycle Sequencing Kit (Amersham Pharmacia Biotech, Inc., Cleveland, USA) and 8% polyacrylamide gel.

Results

We detected somatic mutations only in 12 out of 24 patients with PNH who had been diagnosed by Ham tests. Only two female mutations out of ten female PNH patients were detected. The other 10 mutations were detected out of 14 male patients.

Ten novel mutations and two known mutations were detected (Table 1). Their mutations consisted of five deletions, six substitutions and one insertion. Those mutations resulted in six premature terminations, three abnormal splicings, two nonsense mutations and one missense mutation. The missense mutation was in exon 2. Two patients with venous thrombosis showed mutations in exon 3 only. The patterns of mutations were substitution mutation in six patients and frame shift mutation in the other six. No hot mutation spots were encountered.

PNH III RBCs were 60-93.7% (mean 80.2%) of total RBCs in mutation-positive group and were 9.0-98.7% (mean 46.7%) in mutation-negative group (Table 2). There was difference of mean between mutation-positive and mutation-negative group in PNH III RBCs (P=0.0298).

Discussion

We detected somatic mutations only in 12 out of 24 patients with PNH who had been diagnosed by Ham tests. Such a low detection rate might be most possibly due to leukocytes used for the study. We could separate granulocytes from five recent patients only. Four of them (80%) showed mutations, but 8 of the other 19 (42.1%) did not. The results of male-dominant ddF detection might be another factor. As a normal allele in women even in PNH cells might inhibit detection of mutation, better materials such as granulocytes might be required for higher detection of mutation. Another interesting fact is that CD59 expression for RBCs showed mean differences between mutation-positive and mutation-negative. Although there was inevitable limitation of small number of tests, the more PNH III RBCs were presented, the higher detection rate of mutation was obtained.

As ddF is a convenient method to screen mutations, it may be practical in clinical laboratory. [2] There might be a possibility of lower detection of mutation in 200 bp or larger DNA, [3] but ddF cannot be responsible for the low rate of detection of mutation in this study until other possible factors mentioned above are cleared.

The somatic mutations of the PIG-A gene in Korean patients with PNH varied in site distribution and natures. There was no mutation hot spot. Base substitution and deletion were fifty-fifty in frequency, a frequency nearly equivalent to that reported in a Japanese study [4] but different from a Thai study in which base substitution was found in 10 among 37 mutations in 33 patients. [5] Difference of mutation between ethnic groups may be considered only after enough numbers of mutations are reported.

Two PNH patients with venous thrombosis both showed mutations in exon 3 only. Although lack of correlation between site of mutation and clinical expression is well known, this may be an interesting finding. A missense mutation was found in exon 2 were another interesting finding. There were also more missense mutations in exon 2 than in the other exons in other report. [6] However, a larger sample with a greater number of PNH patients is required to confirm these findings.

References

1. Kim SH, Park SS, Yoon CH, et al.(1993): Analysis of routine test results for the diagnosis of paroxysmal nocturnal hemoglobinuria. Korean J Clin Pathol 13:225-231.
2. Sarkar G, Yoon HS, Sommer SS (1992): Dideoxy Fingerprinting (ddF): A rapid and efficient screen for the presence of mutations. Genomics13:441-443.
3. Nollau P, Wagener C(1997): Methods for detection of point mutations: performance and quality assessment. Clin Chem 43:1114-1128.
4. Yamada N, Miyata T, Maeda K, et al. (1995): Somatic mutations of the PIG-A gene found in Japanese patients with paroxysmal nocturnal hemoglobinuria. Blood 85:885-892.
5. Pramoonjago P, Pakdeesuwan K, Siripanyaphinyo U, et al. (1999): Genotypic, immunophenotypic and clinical features of Thai patients with paroxysmal nocturnal haemoglobinuria Br J Haematol 105:497-504.
6. Nafa K, Bessler M, Castro-Malaspina H et al. (1998): The spectrum of somatic mutations in the PIG-A gene in paroxysmal nocturnal hemoglobinuria includes large deletions and small duplications. Blood Cell Mol Dis 24:370-384.

Table 1. Mutations of the PIG-A gene of patients with PNH

Case No.	Sex/ Age	Site/ Codon	Mutation	Consequence	Suggested site of termination
01	M/25	165-172	22 bases deletion	Premature termination	Codon 194
02	M/20	Intron 4 (5')	G to A	Altered splicing	
04	M/36	Intron 5 (3')	G to A	Altered splicing	
05	M/23	409-417	25 bases deletion	Premature termination	Codon 423
07	F/36	316	ATG→ AG	Premature termination	Codon 329
12	M/39	302	GGA→ TGA	Nonsense	Codon 302
15	M/23	282	GAC→ AC	Premature termination	Codon 290
20	M/52	127	TCA→ TGA	Nonsense	Codon 127
22	M/61	48	GGC→ GTC	Missense (Gly→ Val)	
23	M/47	139	CTC→ C	Premature termination	Codon 171
24	M/37	Intron 5 (5')	G to A	Altered splicing	
25	F/23	248	CCT→ ACCT	Premature termination	Codon 248

Table 2. Results of CD 59 expression of the patients with PNH

Case	CD59 (RBC) (%)			CD59 (Granulocyte) (%)			Mutation
No.	PNH I	PNH II	PNH III	PNH I	PNH II	PNH III	
10	36.8	5.8	57.4	NT	NT	NT	Negative
11	9.0	82.0	9.0	NT	NT	NT	Negative
12	13.0	27.0	60.0	NT	NT	NT	Positive
14	26.0	21.2	52.8	NT	NT	NT	Negative
15	2.8	16.5	80.7	NT	NT	NT	Positive
16	61.0	12.0	27.0	NT	NT	NT	Negative
18	1.3	0.0	98.7	NT	NT	NT	Negative
19	12.0	29.2	58.8	NT	NT	NT	Negative
20	4.6	16.9	78.5	NT	NT	NT	Positive
21	50.8	25.8	23.4	58.5	41.5	0.0	Negative
22	6.3	0.0	93.7	6.4	90.3	3.3	Positive
23	20.3	0.0	79.7	47.0	41.3	11.7	Positive
24	8.3	4.0	87.7	16.2	11.9	71.9	Positive
25	8.0	10.7	81.3	8.0	59.0	33.0	Positive

TWO POSSIBLE MECHANISMS FOR ESCAPE OF GPI⁻ CELLS FROM THE IMMUNOLOGICAL ATTACK

Yoshiko Murakami[1], Hiroshi Kosaka[2], Junji Takeda[3] and Taroh Kinoshita[1]

[1]Department of Immunoregulation, Research Institute for Microbial Diseases, Osaka University
[2]Department of Dermatology, Osaka University Medical School
[3]Department of Environmental Medicine, Osaka University Medical School

Summary: Patients with paroxysmal nocturnal hemoglobinuria (PNH) have one or a few clones of mutant hematopoietic stem cells defective in glycosylphosphatidylinositol (GPI) synthesis due to somatic mutation in the X-linked gene *PIG-A*. The mutant stem cell clone dominates hematopoiesis, the mechanism of which is unclear. To test whether lacks of multiple GPI-anchored proteins result in dysregulation and expansion of stem cells, we generated mice in which GPI-anchor negative cells are present only in hematopoietic system. We transplanted lethally irradiated mice with female fetal liver cells bearing one allele of *Piga* gene disrupted by conditional gene targeting. Due to the X-chromosome inactivation, a significant fraction of the hematopoietic stem cells in fetal livers were GPI-anchor negative. In the transplanted mice, cells of all hematopoietic lineages contained GPI-anchor negative cells. The proportions of GPI-anchor negative cells in various blood cell lineages were stable for 42 weeks, indicating that *Piga* mutation alone does not cause dominance of the mutant stem cells and that other changes are involved in pathogenesis of PNH.

There are two hypothetical mechanisms for clonal expansion of GPI⁻ cells. The one is that the mutant cells are selected by autoreactive cytotoxic cells. The other is that the mutant cell itself gets intrinsic ability to expand by the second hit. Using the previously reported PNH mice in the immunological model system, we studiied to prove the former hypothesis.

Key words: Cre/loxP system, fetal liver transplantation, GPI anchored protein, $CD4^+$T cell, cytotoxic T cell

Introduction

Paroxysmal nocturnal hemoglobinuria (PNH) is an acquired genetic disease in which somatic mutation in *PIG-A* gene is well characterized [1-3], whereas mechanism of clonal expansion has been still unknown [4,5]. It is a hematopoietic stem cell disorder characterized by intravascular hemolysis, venous thrombosis and cytopenia due to bone marrow failure[6,7]. Patients suffering from PNH have abnormal cells in various hematopoietic lineages that are defective in the surface expression of GPI anchored proteins, such as CD59 and CD55. The responsible gene for PNH, X-linked gene *PIG-A*, is involved in the first step of GPI anchor biosynthesis [1].

Although a somatic mutation of *PIG-A* appears to occur in one or a few cells among a large number of hematopoietic stem cells, GPI negative (GPI$^-$) cells dominate over normal population in the bone marrow and the peripheral blood [8]. To test whether *Piga* mutation alone causes the clonal dominancy, we and Rosti and colleagues disrupted mouse *Piga* gene, a homologue of *PIG-A* in the embryonic stem cells and raised chimeric mice bearing GPI$^-$ cells [9,10]. Because GPI-anchored proteins are essential for mouse development [11], mice with high chimerism were not born. To solve this problem of embryonic lethality and to make the chimeric mice bearing GPI$^-$ cells only in hematopoietic lineages , we used Cre/*loxP* system and fetal liver transplantation. Analysis of these chimeric mice demonstrated that percentages of GPI$^-$ cells in all lineages remained constant for several months to one year suggesting that *Piga* mutation alone does not immediately cause the clonal dominance of GPI$^-$ stem cells [12]. Since affected stem cells are defective in the expression of various GPI-anchored proteins and since many GPI-anchored proteins are involved in cell to cell or cell to environment interactions, GPI$^-$ stem cells might escape negative regulation provited by bone marrow environments, resulting in clonal dominance.

One hypothetical mechanism for clonal expansion of GPI$^-$ cells in PNH is that the mutant cells are selected by autoreactive cytotoxic cells [13]. Here we studied it in three model systems.

Methods and Results

First, we made a pair of GPI$^+$ and GPI$^-$ antigen presenting cells from EL4 cell lines that expressed MHC classII and ovalbumin. Lymphocytes from ovalbumin-specific TCR transgenic mice weakly responded to GPI$^-$ cells compared with to GPI$^+$ cells in a proliferation assay.

Second, we transferred a mixture of GPI$^+$ and GPI$^-$ fetal liver cells with or without allogeneic CD4$^+$ T cells into irradiated mice and analyzed the peripheral blood cells. The ratio of GPI$^-$ cells was significantly higher in mice having been transferred with the fetal liver cells together with allogeneic CD4$^+$ T cells.

These results suggest that GPI^- cells are less sensitive to cytotoxic cells because some GPI-anchored proteins work as cofactors.

Third, when a cDNA of GPI-anchored form of ovalbumin is expressed in GPI^+ and GPI^- cells bearing MHC classII, only the GPI^+ cells presented ovalbumin to ovalbumin-specific CD4 T cells. It indicates that if antigen recognized by cytotoxic cells is a GPI-anchored protein, GPI^- cells are less sensitive. Therefore, GPI^- cells can escape immunological attack in two ways; GPI-anchored proteins, working as cofactors and being antigens..

References

1. Miyata T, Takeda J, Iida Y, Yamada N, Inoue N, Takahashi M, Maeda K, Kitani T, Kinoshita T (1993) Cloning of PIG-A, a component in the early step of GPI-anchor biosynthesis. Science 259:1318-1320
2. Takeda J, Miyata T, Kawagoe K, Iida Y, Endo Y, Fujita T, Takahashi M, Kitani T, Kinoshita T (1993) Deficiency of the GPI anchor caused by a somatic mutation of the PIG-A gene in paroxysmal nocturnal hemoglobinuria. Cell 73:703-711
3. Bessler M, Mason PJ, Hillmen P, Miyata T, Yamada N, Takeda J, Luzzatto L, Kinoshita T (1994) Paroxysmal nocturnal haemoglobinuria (PNH) is caused by somatic mutations in the PIG-A gene. Embo J 13:110-117
4. Luzzatto L, Bessler M, Rotoli B (1997) Somatic mutations in paroxysmal nocturnal hemoglobinuria: A blessing in disguise? Cell 88:1-4
5. Young NS (1992) The problem of clonality in aplastic anemia: Dr Dameshek's riddle, restated. Blood 79:1385-1392
6. Rotoli B, Luzzatto L (1989) Paroxysmal nocturnal hemoglobinuria. Semin Hematol 26:201-207
7. Rosse WF (1992) Paroxysmal nocturnal hemoglobinuria. Curr Top Microbiol Immunol 178:163-173
8. Kinoshita T, Inoue N, Takeda J (1995) Defective glycosyl phosphatidylinositol anchor synthesis and paroxysmal nocturnal hemoglobinuria. Adv. Immunol. 60:57-103
9. Kawagoe K, Kitamura D, Okabe M, Taniuchi I, Ikawa M, Watanabe T, Kinoshita T, Takeda J (1996) GPI-anchor deficient mice: Implications for clonal dominance of mutant cells in paroxysmal nocturnal hemoglobinuria. Blood 87:3600-3606
10. Rosti V, Tremmi G, Soares V, Pandolfi PP, Luzzatto L, Bessler M (1997) Murine embryonic stem cells without pig-a gene activity are competent for hematopoiesis with the PNH phenotype but not for clonal expansion. J. Clin. Invest. 100:1028-1036
11. Nozaki M, Ohishi K, Yamada N, Kinoshita T, Nagy A, Takeda J (1999) Developmental abnormalities of glycosylphosphatidylinositol-anchor deficient embryos revealed by Cre/loxP system. Lab. Invest. 79:293-299
12. Murakami Y, Kosaka H, Kinoshita T, Takeda J (1999) Different roles of glycosylphosphatidylinositol in various hematopoietic cells as revealed by model mice of paroxysmal nocturnal hemoglobinuria. Blood 94:2963-2970
13. Young NS, Maciejewski JS (2000) Genetic and environmental effects in paroxysmal nocturnal hemoglobinuria: this little PIG-A goes " Why? Why? Why?" J.Clin. Invest. 106:637-641

Nature of PNH Clones in Aplastic Anemia

Wanachiwanawin W, Piyawattanasakul N, Visuthisakchai S, Issaragrisil S, Department of Medicine, Faculty of Medicine Siriraj Hospital, Mahidol University, Bangkok 10700, Thailand

Summary : Paroxysmal nocturnal hemoglobinuria (PNH) is an acquired clonal, hemolytic disorder characterized by the presence of peripheral blood cells which are deficient in glycosylphosphatidylinositol anchored proteins (GPI – AP). Somatic mutation of the PIG – A gene in a hematopoietic stem cell is responsible for the origin of PNH clone. So far, it is unclear how the mutant stem cell clone dominates hematopoiesis. PNH is closely related to and frequently occurs following aplastic anemia. Our recent study indicated deficient hematopoiesis in vitro in patients with PNH. It is likely that a process inducing bone marrow failure is also responsible for the selection of GPI – AP deficient cells. In order to explore the growth characteristics of PNH clones, we performed regular, repeated flow cytometric analyses of CD59 expression on peripheral blood cells from 32 patients with aplastic anemia. The duration and median time of follow – up were 12-48 and 20 months, respectively. Twenty – two patients (69%) had one or more occurrences of GPI – AP deficient granulocytes, six of them with PNH erythrocytes on the same occasion. Repeated occurrences of PNH cells (up to four times) were observed in 15 patients (47%). Most of the emerging PNH clones were transient and variable in size. Persistence and expansion of the GPI – AP deficient populations to the level of clinical PNH were seen in only four patients (12.5%). As the PNH evolved, GPI – AP deficient granulocytes were more apparent than PNH erythrocytes. Our study indicates that PIG – A mutations of hematopoietic stem cells are relatively common among aplastic anemia patients. Our data argue against the concept of an intrinsic growth advantage of PNH clone, on the contrary they support the hypothesis of a selective process which must be persistently strong enough to favor the outgrowth of a GPI – AP deficient population for the evolution of clinical PNH.

Key words : paroxysmal nocturnal hemoglobinuria, aplastic anemia, glycosyl phospatidylinositol proteins

Introduction

Paroxysmal nocturnal hemoglobinuria (PNH) is an acquired chronic hemolytic disorder characterized by the clonal appearance of peripheral blood cells which are deficient in glycosylphosphatidylinositol anchored proteins (GPI – AP), i.e., PNH cells. Somatic mutation of the X-linked gene, PIG-A in a primitive hematopoietic stem cell is responsible for the origin of a PNH clone [1,2]. It is unclear how the PIG-A mutated stem cells dominate hematopoiesis so that clinical PNH manifests. PNH is closely related to and frequently occurs following aplastic anemia [3]. About 30% of patients who were diagnosed firstly with aplastic anemia subsequently show clinical and laboratory features of PNH [4]. In addition, a recent study has shown deficient hematopoiesis in vitro in patients with PNH, both with (AA – PNH syndrome) or without (de novo PNH) history of aplastic anemia [5]. It is obvious that in a condition of deficient hematopoiesis, PIG-A mutated stem cells have growth advantage over their non-mutated counterparts. A cohort study of the nature of PNH clone in patients with aplastic anemia should provide information and serve as a model for the study of factors favoring growth of PIG-A mutated stem cells.

Patients and Methods

Patients

A total of 32 patients with aplastic anemia (Table 1) were investigated. The diagnosis of aplastic anemia was based on the findings of peripheral blood pancytopenia and hypocellular marrow. There were 15 males and 17 females, with age range of 16 to 66 years with a mean age of 31. The onset of aplastic anemia was between 11 –120 months (median, 60 months) before the study began. The study period was 12 –48 months (median, 20 months) with a 4-month interval between each investigation. All of the patients studied were treated with anabolic hormone and none of them had been transfused within three months prior to the time of each investigation.

Flow cytometric analysis

Flow cytometric analysis of CD 55 and CD 59 of granulocytes and erythrocytes was done according to the method described previously [6]. Granulocytes were purified from heparinized venous blood by sedimentation through high molecular weight dextran and gradient centrifugation through Ficoll-Paques. EDTA blood was used for analysis of the erythrocytes. Granulocytes and erythrocytes were incubated with biotinylated monoclonal antibodies to CD 55 and CD 59 (Pharmingen, CA, USA) and were then labelled with phycoerythrin (PE)-conjugated strepavidin. Mouse IgG 2a was used as an isotypic non-specific background staining control. Since the results from the analysis of CD 55 and CD 59 corresponded, only those of CD 59 are shown. The presence of a PNH clone was recognized when the CD 59 negative population was greater than 1 per cent with a clearcut contour on histogram. The CD 59 positive erythrocytes and granulocytes from 32 normal controls were 99.92 $\pm$ 0.07SD and 99.99 $\pm$ 0.02 SD, respectively.

Ham's acidified serum test

This test was carried out according to the standard method [7].

Statistical analysis

Differences in parameters between the groups were evaluated by the chi-square and the unpaired t test.

Table 1 Patient characteristics and occurrences of PNH blood cells

Number of patients studied	32
Gender (male/female)	15/17
Age in years ; range (median)	16-66 (31)
Duration of study in months ; range (median)	12-48 (20)
PNH granulocytes	
▪ No of patients with one/two or more episodes	7/15
▪ Size of the population as percentage; range (median)	1.0-97.0 (4.3)
PNH erythrocytes	
▪ No of patients with one/two or more episodes	1/5
▪ Size of the population as percentage; range (median)	1.04-16.8 (4.8)

Results

Twenty – two patient (69%) had one or more significant occurrences of CD 59 negative granulocytes, six of them also had PNH erythrocytes on the same occasion (Table 1 and Fig 1) Repeated occurrences of PNH cells (up to four times) were observed in 15 patients (47%). Most of the emerging PNH clones were transient and variable in size (1.0 – 97.0% for granulocytes and 1.04 – 16.8 % for erythrocytes). When together occurred, CD 59 deficient granulocytes were more apparent than PNH erythrocytes.

Among a total of 152 flow cytometric analyses, PNH granulocytes occurred more frequently when the patients' hemoglobin concentrations were below rather than above 12 g/dl (39 episodes of 90 analyses vs 5 episodes of 62 analyses respectively ; $p < 0.005$)

Persistence and expansion of CD 59 negative granulocytes and erythrocytes to the level of a significant positive Ham's test (> 5% lysis) and clinical PNH were observed in only four patients (12.5%).

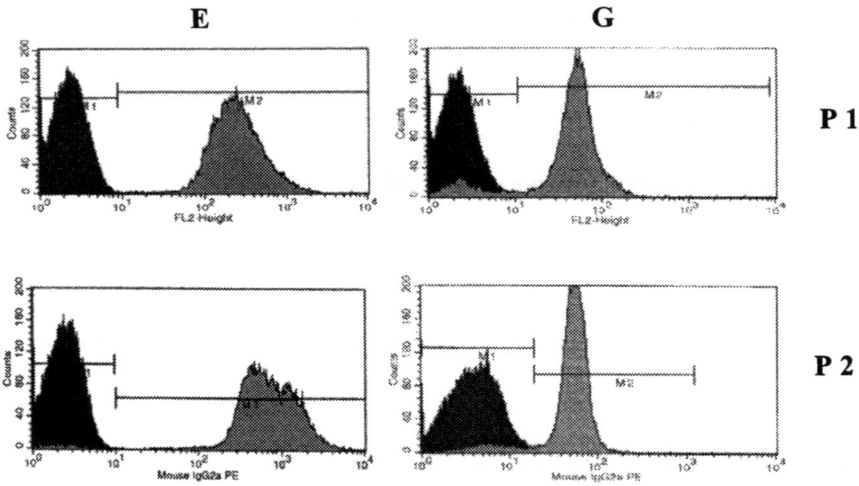

Fig 1 Histogram of CD 59 expression on erythrocytes (E) and granulocytes (G) from two patients with aplastic anemia (P 1 and P 2).

Discussion

According to previous cross-sectional, immunophenotype studies, GPI-AP deficient blood cells have been observed in 23-52% of patients with aplastic anemia [8-10]. In this cohort, flow cytometric analysis of CD 59 on granuloytes and erythrocytes we found that 69% of patients with aplastic anemia demonstrated one or more occurrences of PNH clones. In fact our criteria that a PNH clone requires the presence of at least 1 percent of GPI-AP deficient cell population may underestimate the frequency of the PIG-A mutation in hematopoietic stem cells of aplastic anemia patients. Repeated occurrences of PNH cells in 47% of the patients supports a high incidence of PIG-A mutation in patients with aplastic anemia.

Our study also revealed the behavior of a PNH clone in aplastic anemia. Most of the emerging PNH clones were transient in appearance. Granulocytes were more frequently involved than erythrocytes. Persistence and expansion of the abnormal clone until clinical PNH manifested was observed in only 4 patients (12.5%). During the evolution of PNH, granulocytes appeared to be the first cells affected, followed by erythrocytes and a subsequently positive Ham's test. The abortive nature of most PNH clones in aplastic anemia argues against the intrinsic growth advantage of PIG-A mutant stem cells. On the contrary our study supports the hypothesis of a selective process [11] which must be strong enough to favor the outgrowth of a GPI-AP deficient clone in the evolution of clinical PNH.

Acknowledgement

This work was supported by the Thailand Research Fund (BRG 02/2541)

References

1. Takeda J, Miyata T, Kawagoe K, et al (1993) Deficiency of the GPI anchor caused by a somatic mutation of the PIG-A gene in paroxysmal nocturnal hemoglobinuria. Cell 73 : 703-11
2. Pramoonjago P, Wanachiwanawin W, Chinprasertsuk S, et al (1995) Somatic mutations of PIG-A in Thai patients with paroxysmal nocturnal hemoglobinuria. Blood 86 : 1736-9
3. Kruatrachue M, Wasi P, Na-Kakorn S (1978) Paroxysmal nocturnal haemoglobinuria in Thailand with special reference to an association with aplastic anaemia. Br J Haematol 39 : 267-75
4. Socie G, Mary J Y , De Gramont A, et al (1996) Paroxysmal nocturnal hemoglobinuria : long-term follow up and prognostic factors. Lancet 348 : 573-7
5. Pakdeesuwan K, Muangsup W, Pratya YU, Issaragrisil S, Wanachiwanawin W (2001) Clinical paroxysmal nocturnal hemoglobinuria is the result of expansion of glycosyl- phosphatidyl-insitol-anchored protein-deficient clone in the condition of deficient hematopoiesis. Int J Hematol 73 : 64-70
6. Pakdeesuwan K, Wanachiwanawin W, Siripanyaphingo U, et al (2000) Immunophenotypic discrepancies between granulocytic and erythroid lineages in peripheral blood of patients with paroxysmal nocturnal haemoglobinuria. Eur J Haematol 65 : 8-16
7. Dacie JV, Lewis SM (1984) Practical Haematology, London : Churchill Livingstone, p 201
8. Tooze J, Saso R, March JCW, et al (1995) The novel monolonal antibody By 114 helps detect the early emergence of a paroxysmal nocturnal hemoglobinuria clone in aplastic anemia. Exp Hematol 23 : 1484 – 91
9. Schrezenmeier H, Hertenstein B, Wagner B,. et al (1995) A pathogenetic link between aplastic anemia and paroxysmal nocturnal hemoglobinuria is suggested by a high frequency of aplastic anemia patients with a deficiency of phosphatidylinositol glycan anchored proteins. Exp Hematol 23 : 81-7
10. Yamaguchi M, Nakao S. Takamatsu H, et al (1995) Quality of hematologic recovery in patients with aplastic anemia following cyclosporine therapy. Exp Hematol 23 : 341-6
11. Karadimitris A, Manavalan JS, Thaler HT, et al (2000) Abnormal T-cell repertoire is consistent with immune process underlying the pathogenesis of paroxysmal nocturnal hemoglobinuria. Blood 96 : 2613-20

Serial Analysis of Clonal Expansion in PNH by Flow Cytometry

Jun-ichi Nishimura[1,2], Yuzuru Kanakura[2], Russell E. Ware[1], Tsutomu Shichishima[2], Hideki Nakakuma[2], Haruhiko Ninomiya[2], Sharon Hall[1], Akihisa Kanamaru[2], Hideaki Mizoguchi[2], Mitsuhiro Omine[2], Taroh Kinoshita[2], and Wendell F. Rosse[1]

[1]Duke University Medical Center, Durham NC 27710, USA
[2]Idiopathic Disorders of Hematopoietic Organs Research Committee supported by the Ministry of Health and Welfare of Japan, Tokyo, Japan

Summary: Paroxysmal nocturnal hemoglobinuria (PNH) is an acquired clonal hematopoietic stem cell disorder caused by PIG-A mutations. To assess clonal expansion in PNH over time, patients with serial testing by flow cytometry (interval of at least one year) were analyzed. This study involved 164 patients with PNH from Duke (DP) and 151 patients from Japan (JP). The mean fraction of PMNs deficient in CD59 (mean$\pm$SE) was 75.2$\pm$4.2% in DP and 40.0$\pm$8.3% (n=21) in JP at the initial analysis, and were 74.1$\pm$4.7% and 50.7$\pm$8.6% at the latest analysis (*P*=NS). However, in individual cases, the proportion of affected PMNs varied considerably from -84% to +98%. Some DP and JP pts, however, had a diminution in their CD59-deficient PMNs over time, which was associated with the development of overall hematopoietic failure (*P*=.04 for DP, *P*=.05 for JP). These data illustrate the complexity of clonal expansion in PNH, and the marked variability over time. GPI-deficient populations vary considerably over time for individual patients, but a decreasing PNH clone indicates impending hematopoietic failure. Correlation between clonal expansion or diminution and the development or recovery from aplasia will help our understanding of the natural history of PNH.

Key words: PNH , PIG-A, Flow cytometry, CD59, DAF

Introduction

Paroxysmal nocturnal hemoglobinuria (PNH) is an acquired stem cell disorder characterized by hematopoietic clones with distinct PIG-A mutations and deficient expression of glycosylphosphatidylinositol-anchored proteins (GPI-APs) [1-5].

The lack of GPI-anchored complement regulatory proteins, such as CD55 and CD59, results in complement-mediated hemolysis and hemoglobinuria [1-5]. Flow cytometry of GPI-linked CD55 and CD59 provides an accurate quantitation of the PNH clonal populations [6]. Two large cohorts of patients (pts) with PNH were studied by flow cytometry. To assess clonal expansion in PNH over time, pts with serial testing by flow cytometry (interval of at least one year (yr)) were also analyzed.

Methods and Results

This study involved 164 pts with PNH from Duke (DP) whose diagnosis has been established or confirmed at Duke University, and 151 pts from Japan (JP) from 23 Institutions associated with Idiopathic Disorders of Hematopoietic Organs Research Committee (IDHORC).

The mean age at diagnosis (mean$\pm$SE) was 30.3$\pm$1.4 yrs (range:11-80 yrs) in DP, and 43.6$\pm$1.5 yrs (range:10-82 yrs) in JP ($P<.0001$).

The fractions of cells deficient in CD59 and CD55 at the initial analysis were 45.0 $\pm$2.3% and 39.5$\pm$2.5% in RBCs (P=not significant (NS)), and 71.6$\pm$3.4% and 68.6$\pm$3.3% in PMNs (P=NS) in DP, and 41.1$\pm$3.5% and 37.8$\pm$2.4% in RBCs (P=NS), and 49.6$\pm$4.9% and 42.8$\pm$3.7% in PMNs (P=NS) in JP, respectively. The fraction of RBCs deficient in CD59 at the initial analysis was higher in DP than in JP (P=.03). Similarly, the fraction of PMNs deficient in CD59 was also higher in DP than JP ($P<.0001$). In both patient populations, however, the average proportion of GPI-deficient PMN was higher than GPI-deficient RBCs, which in turn was higher than GPI-deficient mononuclear cells.

Percentages of CD59- and CD55-deficient cells determined by flow cytometric analysis were highly correlated for both RBCs and PMNs (correlation range 0.80-0.95, $P<.0001$). The percentages of affected RBCs and PMNs were also correlated both in CD59- and CD55-deficiencies (correlation range 0.45-0.58, $P<.0008$).

At Duke, 42.1% of pts had intermediate expression (type II) RBCs and 18.6% had Type II PMNs, suggesting that many PNH pts have more than one PIG-A mutant clone.

In both DP and JP, pts who presented with classical symptoms of PNH, including hemoglobinuria, infection or thrombosis, had larger populations of CD59-deficient RBCs and PMNs. Conversely, in both DP and JP, the pts who presented with laboratory evidence of aplasia, including leukopenia or thrombocytopenia, had smaller populations of CD59-deficient RBCs and PMNs.

To assess the change of PNH clone over time, patients who had second testing after an interval of at least one yr were analyzed. Overall, the average fraction of CD59-deficient RBCs was stable over time in both pt populations: 55.3$\pm$4.0% initially and 58.3$\pm$4.3% at followup for DP (P=NS), 39.6$\pm$3.7% initially and 40.5$\pm$4.5% for JP at the latest analysis (P=NS). The average fraction of CD59-

deficient PMNs was also stable over time in both pt populations: 75.2±4.2% initially and 74.1±4.7% at followup for DP (*P*=NS), 40.0±8.3% initially and 50.7±8.6% for JP at the latest analysis (*P*=NS).
Some DP and JP pts, however, had a diminution in their CD59-deficient PMNs over time, which was associated with the development of overall hematopoietic failure (*P*=.04 for DP, *P*=.05 for JP). Other DP and JP pts had an increasing proportion of CD59-deficient PMNs, which was not associated with the development of aplasia.

Conclusions

These data illustrate significant differences between DP and JP with PNH, especially a higher proportion of GPI-deficient cells in Caucasians at the initial analysis. In both populations, however, a larger PNH clone was associated with classical PNH symptoms, while a smaller PNH clone was associated with aplasia. GPI-deficient populations vary considerably over time for individual patients, but a decreasing PNH clone indicates impending hematopoietic failure.
These data illustrate the complexity of clonal expansion in PNH, and the marked variability over time. Correlation between clonal expansion or diminution and the development or recovery from aplasia will help our understanding of the natural history of PNH.

Acknowledgments

This work was partly supported by a grant from The Japan Intractable Diseases Research Foundation and The Osaka Medical Research Foundation for Incurable Diseases.
We gratefully thank Drs. T. Uchiyama, K. Tohyama, A. Urabe, K. Ozawa, M. Tomonaga, H. Hirai, K. Mitani, T. Murate, Y. Niho, T. Ueda, R. Ohno, K. Ohyashiki, M. Karasawa, A. Kimura, M. Teramura, Y. Niitsu, M. Harada, M. Bessyo, T. Hotta and S. Nakao for providing patients data.
We also thank T. Hirota and Y. Murakami for their excellent technical assistance.

References

1. Nishimura J, Murakami Y, Kinoshita T (1999) Paroxysmal nocturnal haemoglobinuria:An Acquired Genetic Disease. Am J Haematol 62:175-182
2. Ware RE, Hall SE, Rosse WF (1991) Paroxysmal nocturnal hemoglobinuria with onset in childhood and adolescence. N Engl J Med 325:991-996

The Clinical Course of PNH in the USA and in JAPAN

Jun-ichi Nishimura[1,2], Yuzuru Kanakura[2], Russell E. Ware[1], Tsutomu Shichishima[2], Hideki Nakakuma[2], Haruhiko Ninomiya[2], Sharon Hall[1], Akihisa Kanamaru[2], Hideaki Mizoguchi[2], Mitsuhiro Omine[2], Taroh Kinoshita[2], and Wendell F. Rosse[1]

[1]Duke University Medical Center, Durham NC 27710, USA
[2]Idiopathic Disorders of Hematopoietic Organs Research Committee supported by the Ministry of Health and Welfare of Japan, Tokyo, Japan

Summary: PNH, which is characterized by intravascular hemolysis and venous thrombosis, is an uncommon disorder of hematopoietic stem cells with an acquired somatic mutation of PIG-A. An understanding of the natural history of PNH is essential to improve therapy. To determine and to directly compare the clinical courses of PNH patients from the USA and Japan, data were collected for 385 patients with PNH. There were 176 patients from Duke (DP) and 209 patients from Japan (JP). The mean age at diagnosis (mean$\pm$SE) was 32.8$\pm$1.2 years (4-80) in DP, and 45.1$\pm$1.3 years (10-86) in JP. The incidence of various clinical events was analyzed. These analyses provide important diagnostic and long-term data on two large cohorts of PNH patients, and identify important differences between Caucasian and Asian patients with PNH. Caucasians tend to be younger and have classical symptoms of PNH including thrombosis, while Asians tend to be older and have symptoms of aplasia. Multivariate analysis of factors influencing survival was also undertaken. By this study, estimates of PNH prognostic factors may provide a better understanding of the complications of the disease so that therapeutic interventions may be sought. Further, by noting differences in the two populations, possible genetic modifiers of the course of the disease may be identified for further investigation.

Key words: PNH , PIG-A, Aplastic anemia, Thrombosis, Hemoglobinuria

Introduction

Paroxysmal nocturnal hemoglobinuria (PNH) is an uncommon acquired hematopoietic stem cell disorder characterized by intravascular hemolysis and venous thrombosis [1-5]. The lack of glycosylphosphatidylinositol (GPI)-anchored complement regulatory proteins, such as CD55 and CD59, results in complement-mediated hemolysis and hemoglobinuria [1-5]. In the affected hematopoietic cells from patients with PNH, the first step in biosynthesis of the GPI anchor is defective [1]. The X-linked gene termed PIG-A, the product of which participates in this reaction step, has been found to be mutated in populations of the hematopoietic cells of all patients with PNH reported to date [1]. An understanding of the natural history of PNH is essential to improve therapy [2-5]. To determine and to directly compare the clinical courses of PNH patients from the USA and Japan, data were collected for 385 patients with PNH.

Methods and Results

This study involved 176 patients from Duke (DP) who were mostly Caucasian, and 209 patients from Japan (JP) who were exclusively Asian. The average age at diagnosis (mean ± SE) was younger for DP (32.8 ± 1.2 years, range 4-80 years), compared to JP (45.1 ± 1.3 years, range 10-86 years), P<.0001.

There were more female DP (56.3%) than female JP (43.5%), *P*=.01. At the time of diagnosis, DP had a lower incidence of previous aplastic anemia than JP (29.0% vs 37.8%, *P*=.07), but a similar incidence of previous MDS (5.1% vs 4.8%, *P*=NS). DP had a higher incidence of classical PNH symptoms than JP at diagnosis, including hemoglobinuria (50.0% vs 33.5%, *P*=.001), thrombosis (19.3% vs 6.2%, *P*<.0001), and infection (13.6% vs 3.4%, *P*=.0002). In contrast, DP had a lower incidence at diagnosis of symptoms from anemia (88.1% vs 94.3%, *P*=.03), neutropenia (45.5% vs 72.3%, *P*<.0001), and thrombocytopenia (52.3% vs 63.2%, *P*=.03).

At diagnosis, DP had a higher hemoglobin concentration (9.7 gm/dL vs 8.2 gm/dL, *P*<.0001), higher ANC (3.0 x 10^9/L vs 1.8 x 10^9/L, *P*<.0001), higher platelet count (140 x 10^9/L vs 96 x 10^9/L, *P*<.0001), higher reticulocyte count (195 x 10^9/L vs 78 x 10^9/L, *P*<.0001), and higher LDH (2337 vs 1572 IU/L, *P*=.1).

During follow-up, more DP than JP developed a new thrombotic event (31.8% vs 4.3%, *P*<.0001) and severe infection (18.2% vs 9.1%, *P*=.009), but the number of patients with hematopoietic failure (33.0% vs 36.4%, *P*=NS), MDS (3.4% vs 3.8%, *P*=NS), and leukemia (0.6% vs 2.9%, *P*=NS) was not significantly different.

During follow-up, more JP than DP did not require any treatment (28.2% vs 5.7%, *P*<.0001) (Table 4). In contrast, more DP than JP required transfusion (61.9% vs 40.2%, *P*<.0001), Prednisone (82.4% vs 46.9%, *P*<.0001), ATG/ALG (15.3% vs

2.9%, $P<.0001$), anti-coagulant (26.7% vs 4.3%, $P<.0001$), and BMT (8.0% vs 1.9%, $P=0052$).

Surprisingly, 6 babies were born from 5 DP and 14 babies from 8 JP, and 4 maternal DP and 1 JP had complications of thrombosis. At the latest follow-up, 4 DP (2.3%) and 10 JP (4.8%) developed spontaneous remission (P=NS) (Table 5). Survival analysis revealed a similar number of deaths in each group: 21.2% of DP vs 21.8% of JP.

However, the causes of death were different (for example, 41.7% of DP deaths were from thrombosis, compared to 7.9% of JP deaths, $P=.0007$).

JP had a longer mean survival time than DP (385 months vs 234 months), although the Kaplan-Meier survival curves were not significantly different.

Conclusions

These analyses provide important diagnostic and long-term data on two large cohorts of PNH patients, and identify important differences between Caucasian and Asian patients with PNH. Caucasians tend to be younger and have classical symptoms of PNH including thrombosis, while Asians tend to be older and have symptoms of aplasia. Possible genetic modifiers of the course of PNH may be identified with further investigation.

Acknowledgments

This work was partly supported by a grant from The Japan Intractable Diseases Research Foundation and The Osaka Medical Research Foundation for Incurable Diseases.

We gratefully thank Drs. T. Uchiyama, K. Tohyama, A. Urabe, K. Ozawa, M. Tomonaga, H. Hirai, K. Mitani, T. Murate, Y. Niho, T. Ueda, R. Ohno, K. Ohyashiki, M. Karasawa, A. Kimura, M. Teramura, Y. Niitsu, M. Harada, M. Bessyo, T. Hotta and S. Nakao for providing patients data.

We also thank T. Hirota and Y. Murakami for their excellent technical assistance.

References

1. Nishimura J, Murakami Y, Kinoshita T (1999) Paroxysmal nocturnal haemoglobinuria:An Acquired Genetic Disease. Am J Haematol 62:175-182
2. Ware RE, Hall SE, Rosse WF (1991) Paroxysmal nocturnal hemoglobinuria with onset in childhood and adolescence. N Engl J Med 325:991-996

Long-Term Support of Human Hematopoiesis by a Single Stem Cell Clone

Toshiyuki Hirota[1,2,3], Jun-ichi Nishimura[1,2], Yuzuru Kanakura[2], Takashi Machii[2], Maki Kuwayama[1,2], Norimitsu Inoue[1], Kazuhito Ohishi[1], and Taroh Kinoshita[1]

[1]Department of Immunoregulation, Research Institute for Microbial Diseases, Osaka University, Suita, Osaka 565-0871, Japan
[2]Department of Hematology and Oncology, Osaka University Medical School, Suita, Osaka 565-0871, Japan
[3]Research Foundation for Microbial Diseases of Osaka University, Suita, Osaka 565-0871, Japan

Summary: Paroxysmal nocturnal hemoglobinuria (PNH) is an acquired clonal hematopoietic stem cell disorder caused by PIG-A mutations. Many patients with PNH have more than one PNH clone, but it is unclear whether a single PNH clone remains dominant or if minor clones eventually become dominant. Furthermore, it is unknown how many hematopoietic stem cells (HSCs) sustain hematopoiesis and how long a single HSC can support hematopoiesis in humans. To understand the dynamics of HSCs, we reanalyzed the PIG-A gene mutations in 9 patients with PNH 6 to 10 years after their initial analysis. The proportions of affected peripheral blood PMNs in individual patients were highly variable; 2 patients increased, 3 patients decreased, and 4 had stable populations (changed less than 20%). In all cases, the previously predominant clone was still dominant and present as assessed by DNA analysis. Two patients whose affected PMNs decreased and who have been followed significantly longer than others, now have an aplastic condition, suggesting that aplasia is a terminal feature of PNH. This is the first report directly demonstrating that one stem cell clone can sustain human hematopoiesis for 6 to 10 years. The complexity of these results reflects the marked variability of the clinical course of PNH.

Key words: PNH, PIG-A mutant clone, Hematopoietic stem cell, Clonal dominance, Bone marrow aplasia

Introduction

Paroxysmal nocturnal hemoglobinuria (PNH) is an acquired hematopoietic stem cell disorder characterized by clonal blood cells that are deficient in glycosylphosphatidylinositol-anchored proteins (GPI-APs) due to somatic mutations of the PIG-A gene [1,2]. Many patients with PNH have more than one PNH clone [1,3]. How the affected stem cell clone comes to dominate hematopoiesis is a current issue. It is also unclear whether the predominant clone is maintained or a minor clone eventually overcomes. It is not known how many hematopoietic stem cells (HSCs) sustain hematopoiesis and how long one HSC could support hematopoiesis in human. To understand dynamics of HSCs, we reanalyzed the PIG-A gene mutations in 9 patients 6 to 10 years after the previous analyses [2-4].

Methods and Results

Changes in proportion of GPI-AP$^-$ cells

To investigate the pattern of clonal dominance, we reanalyzed the surface expression of GPI-APs on PBPMNs from 9 patients with PNH 6-10 years after the previous analysis. We described all the patients elsewhere [2,3,4]. The mean percentage of PBPMNs deficient in CD59 (a measure of clonal hematopoiesis) was 83.4±5.6% at the previous analyses and 70.8±7.9% at the current analysis (difference not significant, Wilcoxon signed rank test). However, a change in the proportion of CD59 deficient PMNs was highly variable from patient to patient; it increased in 2 (50% to 98% in J4 and 65% to 97% in J5), was stable in 4 (97% to 93% in J3, 83% to 82% in J11, 87% to 73% in J13 and 78% to 61% in J19) and diminished in 3 patients (94% to 33% in J12, 99% to 57% in J15 and 98% to 43% in J16).

We also investigated the absolute numbers of CD59 negative PMNs in each patient. In the groups with an increased (J4 and J5) and a diminished (J12, J15 and J16), the absolute numbers of affected PMNs were increased and decreased, respectively. Whereas, in the group with a stable, the numbers of affected PMNs were stable in J3 and J11, and were decreased in J13 and J19, whose proportions of affected PMNs were stable due to a decline of both PNH and normal clones.

Changes in proportion of PIG-A mutant clones

We then analyzed PIG-A gene in the PBPMNs after amplification and subcloning of appropriate regions. In all cases, the previously predominant mutant clone was still dominant. A proportion of affected PMNs in patient J5 increased due to the expansion of the predominant clone. In patient J4, whose proportion of affected PMNs increased from 50% to 98%, the predominant clone was detected in 3 of 10 amplified subclones (AS) of DNA and the second clone was newly detected in 2 of 10 AS. Since this patient is female, this result indicates that two mutant clones almost completely occupy the hematopoiesis currently and that the first clone supported about a half of hematopoiesis for 7 years. In 4 patients with stable proportion of affected PMNs (J3, J11, J13 and J19), the dominant clones are still dominant 7 to 8 years later. Blood samples from J19, bearing 4 independent PIG-A mutant clones [3], were analyzed in detail. Sixty percent of PBPMNs and ninety percent of BMMNCs were defective in CD59 expression. The predominant clone was detected in 5 of 10 AS from PBPMNs and 10 of 10 AS from BMMNCs, indicating that proportions of abnormal phenotype and genotype are well correlated. A minor type mutation was only detected in 1 of 16 AS from BMMNCs, and 2 other minor mutations and any additional mutations were not detected. These results suggest that the predominant clone might have superceded the other 3 minor clones; alternatively these minor clones may have declined spontaneously. Proportions of affected PMNs in 3 patients (J12, J15 and J16) decreased due to a decline of the major PNH clone. In patient J12, whose predominant clone, previously detected in 14 of 20 AS, was detected in only 4 of 16 AS; a minor clone, previously detected in 1 of 12 AS, was found in 2 of 11 AS. In patient J15, whose predominant clone, previously detected as dominant, was detected in only 4 of 10 AS; a minor clone at previous study was found in 1 of 9 AS. Thus, the minor clones in 2 patients (J12 and J15) remained as minor and the decreases of affected PMNs were due to decline of the predominant clones.

Conclusions

In all cases, the previously predominant clone was still dominant, indicating that the predominant PNH clone mainly supported hematopoiesis and manifested of disease for years. However, the minor clone in 2 patients (J4 and J12) is now competing with the previously predominant clone, and may overcome in the future.

Finally, we also investigated the clinical course especially in 3 patients (J12, J15 and J16), whose affected PMNs decreased in proportion, contrary to our expectation, in spite of their growth/survival advantage. Interestingly, 2 patients (J15 and J16) out of 3 eventually have developed to an aplasia from de novo PNH, and their present condition is clinically poor despite the decrease in affected PMNs. In addition, these 2 cases have been followed significantly longer than other patients, although there is no significant disparity in age between these groups. Thus, aplasia may be a terminal feature of PNH, as the abnormal clone declines in hematopoietic capacity. Active hematopoiesis of PNH stem cell clone together with desperate hemolysis is rather better than aplasia due to a decline of PNH stem cell clone.[5] Therefore, our observations suggest that immunosupressive therapy, such as anti-thymocyte globulin or anti-lymphocyte globulin, may be effective in these cases.

Acknowledgments

This work was partly supported by a grant from The Japan Intractable Diseases Research Foundation and The Osaka Medical Research Foundation for Incurable Diseases.

References

1. Nishimura J, Murakami Y, Kinoshita T (1999) Paroxysmal nocturnal haemoglobinuria:An Acquired Genetic Disease. Am J Haematol 62:175-182
2. Miyata T, Yamada N, Iida Y, et al (1994) Abnormalities of PIG-A transcripts in granulocytes from patients with paroxysmal nocturnal hemoglobinuria. N Engl J Med 330:249-255
3. Nishimura J, Inoue N, Wada H, et al (1997) A patient with paroxysmal nocturnal hemoglobinuria bearing four independent PIG-A mutant clones. Blood 89:3470-3476
4. Ueda E, Nishimura J, Kitani T, et al (1992) Deficient surface expression of glycosylphosphatidylinositol-anchored proteins in B cell lines established from patients with paroxysmal nocturnal hemoglobinuria. Int Immunol 4:1263-1271

Inflammatory defects caused by GPI-anchor deficiency in macrophages

Wouter Hazenbos[1], Irmgard Förster[2], Björn Clausen[2], Junji Takeda[3] and Taroh Kinoshita[1]

[1]Dept. of Immunoregulation, Research Institute for Microbial Diseases, Osaka University, 3-1 Yamada-oka, Suita, Osaka 565-0871, Japan; [2]Institute for Med. Microbiology, Immunology and Hygiene, Technical University Munich, Trogerstr. 4b, D-81675 Munich, Germany; and [3]Dept. of Social Environmental Medicine, Osaka University Medical School, 2-2 Yamadaoka, Suita, Osaka 565-0871, Japan

Summary: The function of the glycosyl phosphatidylinositol (GPI)-anchor on macrophages was studied using myeloid-specific GPI-anchor knock-out mice, which were generated by conditional targeting of *PigA* in the myeloid lineage. Alveolar macrophages from these mice displayed a GPI-anchor negative phenotype, as determined by the lack of heat stable antigen and CD48 on their surface. These macrophages did not produce the proinflammatory cytokine tumor necrosis factor (TNF)-α in response to a low concentration of lipopolysaccharide in vitro. After induction of immune complex-mediated alveolitis, an in vivo model for inflammation, several inflammatory processes were impaired in these mice, including local production of the macrophage-derived proinflammatory cytokines TNF-α and interleukin-6, subsequent neutrophil infiltration, and hemorrhage. Together, these observations indicate that the GPI-anchor on macrophages serves a nonredundant stimulatory function in inflammation, triggered either through the innate or the acquired immune system.

Key words: GPI-anchor, inflammation, macrophages, immune complexes, TNF-α

Introduction

Leukocytes express a variety of surface proteins that are linked to the cell membrane through the GPI-anchor. Several studies, the majority of which being performed in vitro, suggest that GPI-anchored proteins participate in various leukocyte functions. In the hematological disorder paroxysmal nocturnal hemoglobinuria (PNH), the GPI-anchor is absent on varying numbers of all types of leukocytes, due to a somatic mutation in the *PigA* gene [1, 2], which is essential for the biosynthesis of the GPI anchor [3], in hematopoeitic stem cells. One of the main causes of death in PNH patients is infection, raising the possibility that GPI anchor-deficiency causes a loss of certain vital functions of the immune system. Little is known about the functions of the GPI-anchor on macrophages, which are major effector cells in inflammation and protection against infection. The purpose of this study was to investigate the function of the GPI-anchor on macrophages using a myeloid-specific GPI-anchor knock-out mouse.

Methods

Myeloid-specific GPI-anchor knock-out mice were generated by using Cre/LoxP conditional targeting, in which LoxP-flanked *PigA* [4] was targeted through myeloid-specific expression of Cre recombinase [5]. Alveolar macrophages were isolated by bronchoalveolar lavage. Expression of GPI-anchored proteins was analyzed using flow cytometry. Macrophages were cultured in vitro with lipopolysaccharide, and the release of tumor necrosis factor (TNF-α)into the culture supernatant was measured. Immune complex-mediated alveolitis was induced by intratracheal injection of the mice with IgG anti-ovalbumin and intravenous injection with ovalbumin. To determine the extent of inflammation, concentrations of TNF-α and interleukin-6, numbers of infiltrating neutrophils, and numbers of erythrocytes as a measure for hemorrhage, in the bronchoalveolar fluids of these mice were measured.

Results

Alveolar macrophages from these mice displayed a GPI-anchor negative phenotype, as determined by the lack of heat stable antigen and CD48 on their surface. These macrophages did not produce the proinflammatory cytokine TNF-α in response to low concentrations of lipopolysaccharide in vitro. After induction of immune complex-mediated alveolitis, an in vivo model for inflammation, several inflammatory parameters were impaired in these mice, including local production of the macrophage-derived proinflammatory cytokines TNF-α and interleukin-6, subsequent neutrophil infiltration, and hemorrhage.

Conclusion

Thus, GPI anchor-deficiency in macrophages causes impaired responses to different inflammatory stimuli in vitro and in vivo. This indicates that GPI-anchored proteins on macrophages have a nonredundant stimulatory function in inflammation. The observed decreased response of GPI anchor-deficient macrophages to lipopolysaccharide is consistent with previous studies using blood from PNH patients, and is likely due to a loss of the GPI-anchored protein CD14. This defective inflammatory response to lipopolysaccharide indicates a role for the GPI-anchor in triggering of the innate immune system. The present study also provides evidence that the GPI-anchor contributes to immune complex-mediated inflammation. This indicates a previously unrecognized function for the GPI-anchor in inflammation triggered through the acquired immune system.

References

1. Takeda J, Miyata T, Kawagoe K et al (1993) Deficiency of the GPI anchor caused by a somatic mutation of the *PIG-A* gene in in paroxysmal nocturnal hemoglobulinuria. Cell 73: 703-711
2. Kinoshita T, Inoue N, and Takeda J (1996) Role of phosphatidylinositol-linked proteins in paroxysmal nocturnal hemoglobulinuria pathogenesis. Annu Rev Med 47:1-10
3. Miyata T, Takeda J, Iida Y, et al (1993) The cloning of PIG-A, a component in the early step of GPI-anchor synthesis. Science 259:1318-1320
4. Tarutani M, Itami S, Okabe M, et al (1997) Tissue-specific knockout of the mouse Pig-a gene reveals important roles for GPI-anchored proteins in skin development. Proc Natl Acad Sci USA 94:7400-7405
5. Clausen BE, Burkhardt C, Reith W, et al (1991) Conditional gene targeting in macrophages and granulocytes using LysMcre mice. Transg Res 8:265-277

Paroxysmal Nocturnal Hemoglobinuria in Association with Chronic Myelofibrosis

Hiroshi Harada, Hiraku Mori, Haruo Niikura and Mitsuhiro Omine

Division of Hematology, Internal Medicine, Showa University Fujigaoka Hospital
1-30 Fujigaoka, Aoba-ku, Yokohama 217-8501, Japan

Summary: Two cases of chronic idiopathic myelofibrosis (CIMF) associated with paroxixmal nocturnal hemoglobinuria (PNH) are descrived. Case 1 was 60 yrs old male. He was diagnosed CIMF in 1990. Association of PNH was diagnosed in 1995. PNH clone expand gradually, frequent hemolytic episodes were observed. He died by liver failure in 1999. Case 2 was 51 yrs old male. He was daiagnosed as CIMF in 1993 and association of PNH was diagnosed in 2000. Case 2 has no typical hemolytic episode so far. Since both diseases belong to clonal diorder, it is suggested clonal evolution was occurred in CIMF and developed PNH.

Key words: paroxysmal nocturnal hemoglobinuria, chronic idiopathic myelofibrosis, neutorophil alkaline phosphatase stain, clonal evolution

Introduction

Chronic idiopathic myelofibrosis (CIMF) is a progressive bone marrow disorder which characterized by 1) diffuse fibrosis and sclerosis of bone marrow, 2) huge hepatosplenomegaly due to extramedullary hematopoiesis, 3) leukoerythroblastic anemia. The etiology of CIMF is still unknown, but clonal proliferation of an abnormal hematopoietic multipotent stem cell was observed.
Paroxysmal nocturnal hemoglobinuria (PNH) is acquired hemolytic anemia due to expansion of clone which lack GPI anchor protein. It is known that PNH occasionally complicated to other hematological disorders as well as with aplastic anemia. But the reports of CIMF, which complicated PNH, are limited. We report here two cases of PNH which were developed during their CIMF courses.

Case report

Case 1:

In August 1990, 60 years old male was introduced our hospital because of high LDH level and blast in peripheral blood during follow up of postgastrectomy due to gastric ulcer. He had huge liver (6 cm below costal margin) and spleen (10 cm below costal margin) and his laboratory data showed leukoerythroblastic anemia. Bone marrow biopsy revealed the presence of chronic myelofibrosis (Fig. 1A.).

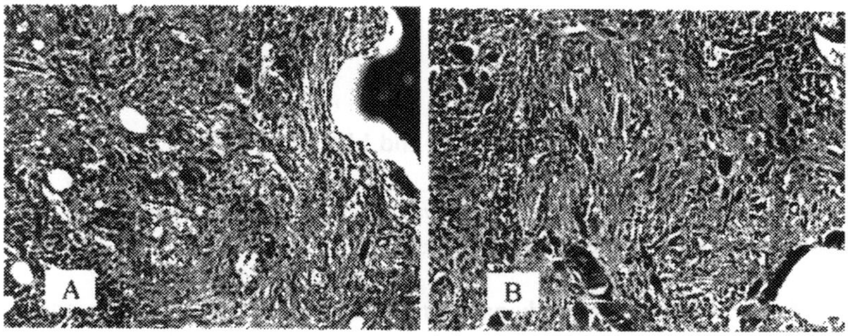

Fig. 1A, B.
Bone marrow biopsies of case 1 (A) and case 2 (B) show typical myelofibrosis. (HE stain, original magnification is x 200.)

Testosterone therapy was started. In April 1993, frequent RBC transfusion was started because of anemia progression. In July 1993, his neutrophil alkaline phosphatase (NAP) positivity was 89%, and negative for Ham test and sugar-water test was confirmed. Buslufan pulse therapy was started. In December 1995, NAP positivity was declined to 39%. Also Ham test and sugar-water test were converted into positive. The presence of $CD55^-/CD59^-$ clone was detected by flow cytometry method (Fig. 2D, A, B, C.). He was diagnosed as PNH in association with CIMF. In October 1998, splenic irradiation was performed for progressive splenomegaly. At this time, chromosomal abnormality (46, XY, del(5)(q?31), add(18)(q23)) was detected from peripheral blood. The percentage of $CD55^-/CD59^-$ clone in peripheral blood was increased time dependently (Fig. 2D, E, A.), and most $CD34^+$ cells were converted to $CD55^-/CD59^-$ clone (Fig. 2F.). Ever since then, frequent hemolysis was observed. He died by liver failure in July 1999.

Case 2:

51 years old male was consulted local hospital due to abdominal fullness in January 1993. He had huge liver and spleen (6 cm below costal margin). Peripheral blood analysis revealed he had leukocytosis, thrombocytosis, and mild anemia. His NAP positivity was 98%. Massive bone marrow fibrosis was

observed by trephine bone marrow biopsy. He was diagnosed as CIMF. Hydroxycarbamide was prescribed. In February 1999, his splenomegaly and anemia progressed. He needed occasional RBC transfusion. He was transferred to our hospital due to progressive anemia and heart failure in June 2000. He had huge liver and spleen (6 cm and 16 cm below costal margin, respectively). It was reconfirmed as myelofibrosis by bone marrow biopsy (Fig. 1B.). Because of the decline of NAP positivity (54%), $CD55^-/CD59^-$ clone was checked by flow cytometry method. PNH clone was detected from peripheral blood (Fig. 2G to J.), and no chromosomal abnormality was detected. Although Ham test was negative, sugar water test was positive. The diagnosis of PNH with CIMF was made. After busulfan pulse therapy was performed for several times, his anemia was improved slightly. No typical hemolytic episode was observed till June 2001.

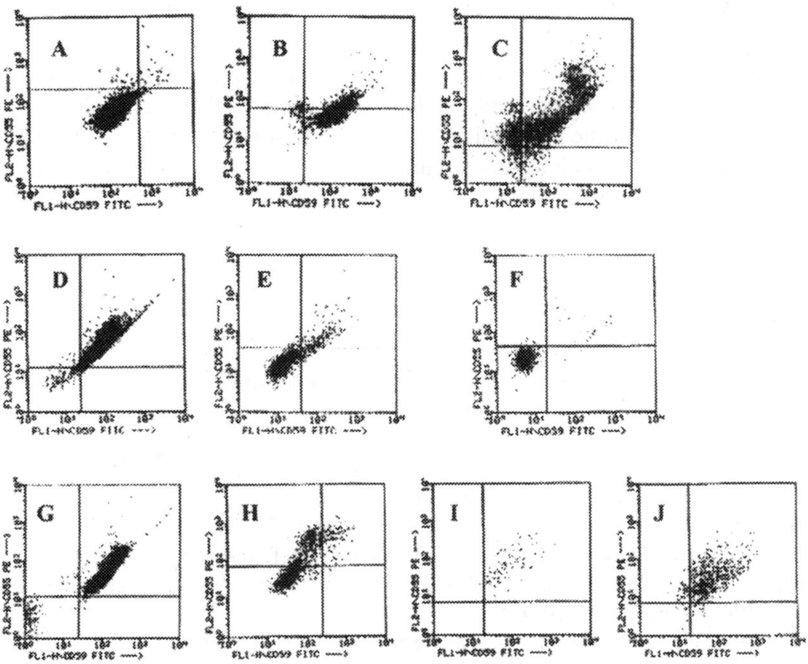

Fig. 2A to J.

Flow cytemetlic analysis of CD55 (vertical axis) and CD59 (horizontal axis) expression in peripheral blood cells in case 1 (A to F) and case 2 (G to J). Samples of A to C were obtained in May 1999, sample of D was in Dec. 1995, and samples of E and F were in Dec. 1998. Samples of G to J were obtained in Jul. 2000. Each samples were fractionated to erythrocyte (G), granulocyte (A, D, E, H), monocyte (B, I), lymphocyte (C, J) and CD34+ cells (F), then analyzed for CD55 and CD59 expression.

Discussion

PNH is a hematological clonal disorder anemia which is caused by a somatic mutation of the PIG-A gene [1]. CIMF is one of the chronic myeloproliferative disorders and also belong to hematological clonal disorder [2].

We report here two cases of PNH which is associated with CIMF. Both cases were diagnosed and followed up as CIMF first (Fig. 1A, B.). A numbers of years later, they developed PNH. The PNH clones were identified in the most of blood cells in various rate (Fig. 2A to J.), and expanded from lower rate to higher rate with time (Fig. 2D, E, A.). In $CD34^+$ cells, PNH clone was accounted for the majority part (Fig. 2F).

It is well recognized that some patients with aplastic anemia develop PNH. But the cases of PNH which associated with CIMF were rearly reported [3-8]. The association PNH with CIMF was first reported by Damshek et al [3]. Hansen et al found two patients who had PNH signs and symptoms, and one patient who was positive for suc in ten myelofibrosis patients [4]. Chi-yen et al reported 54% of primary myelofibrosis patients had "PNH-like defects" of the red blood cell membrane [5]. Althoug the mechanism of the association of PNH to CIMF is not clear, since both diseases belong to clonal disorder, these cases suggest the clonal evolution develop PNH from CIMF.

References

1. Takeda J, Miyata T, Kawagoe K et al (1993) Deficiency of the GPI anchor caused by a somatic mutation of the PIG-A gene in paroxysmal nocturnal hemoglobinuria. Cell 73:703-711.
2. Lebowitz P, Papac R, Ghosh PK (1990) Impaired retinoblastoma susceptibility (Rb) gene in agnogenic myeloid metaplasia. Blood (suppl. 1) 76:236a.
3. Damshek W, Fudenberg H (1957) Paroxysmal nocturnal hemoglobinuria, atypical manifestations suggesting an immunologic disease. Arch Int Med 99:202-208
4. Hansen NB, Killmann SA (1970) Paroxysmal Nocturnal Hemoglubinuria in Myelofibrosis. Blood 36:428-431.
5. Chi-yen Kuo, Van Voolen G, Morrison AN (1972) Primary and Secondary Myelofibrosis: Its Relationship to "PNH-like Defect". Blood 40:875-880.
6. Lewis SM, Pettit JE, Tattersall MHN et al (1971) Myelosclerosis and Paroxysmal Nocturnal Haemoglobinuria. Scand J Haemat 8:451-460.
7. Hirai H, Asahara A, Suzuki R (1973) A Case of Myelofibrosis with Paroxysmal Nocturnal Hemoglobinuria (in Japanese). Rinsho Ketsueki 14:805-811.
8. Almquist A, Wahlin A (1981) Response to Busulfan Treatment of Paroxysmal Nocturnal Hemoglobinuria and Myelofibrosis in One and the Same Patient. Acta med Scand 209:133-135.

Glycosyl-Phosphatidyl-Inositol anchored protein-deficient blood cells in children with aplastic anemia

Ayami Yoshimi, Yoshiyuki Takahashi, Kazuko Kudo, and Seiji Kojima

The Department of Pediatrics and Developmental Pediatrics, Nagoya University Graduate School of Medicine

Summary: The deficiency of Glycosyl-Phosphatidyl-Inositol anchored protein (GPI-AP) on granulocytes and red blood cells (RBCs) was studied using flow cytometry in 30 children with aplastic anemia (AA); 10 of whom were yet to receive immunosuppressive therapy (pre-IST group) and 20 of whom had undergone IST treatment (post-IST group). None of the patients had clinical symptoms suggesting paroxysmal nocturnal hemoglobinuria (PNH). The GPI-AP–deficient populations in granulocytes were 0.02±0.041, 0.005±0.012, and 0.001±0.001%, in the pre-IST group, post-IST group, and healthy controls, respectively. In RBCs, the equivalent figures were 0.194±0.469, 0.018±0.027, and 0.002±0.003%, respectively. Only one patient in the pre-IST group had more than 1% of GPI-AP deficient populations in RBCs. Our results showed that the frequency of GPI-AP–deficient blood cells is low in children with AA. This is consistent with the clinical finding that few AA children develop PNH during or following IST.

Key words: Glycosyl-Phosphatidyl-Inositol anchored protein, aplastic anemia, children, paroxysmal nocturnal hemoglobinuria

Intoduction

Although several studies using flow cytometry have shown that circulating blood cells with a deficiency of Glycosyl-Phosphatidyl-Inositol anchored protein (GPI-AP) are detected in up to 51% of adult aplastic anemia (AA) patients

[1][2][3], there has been no equivalent investigation for children with AA. We, therefore, examined GPI-AP expression on peripheral blood cells (PBCs) in 30 children with AA.

Patients and methods

Subjects were 10 children with AA newly diagnosed between 4/2001 and 7/2001 and yet to receive immunosuppressive therapy (pre-IST group) and 20 children who had previously been diagnosed with AA and received immunosuppressive therapy (post-IST group). The patients in the latter group had undergone immunosuppressive therapy with antithymocyteglobulin and cyclosporin A up to 10 years before this study, and had received no red blood cell (RBC) transfusions for at least 6 months prior to enrollment in this study. Blood samples were collected from all 30 children and from 10 healthy adults (control group). One child with symptomatic AA-PNH syndrome was also examined. Granulocytes were isolated from heparinized peripheral blood, triply stained with phycoerythrin (PE)-conjugated anti-CD55 MoAb (Serotec, Oxforf, England), PE-conjugated anti-CD59 MoAb (Pharmingen, San Diego, CA), and fluorescein-isothiocyanate (FITC)-conjugated anti-CD 13 MoAb (Immunotech, Marseille, France), and analyzed by FACScan (Becton Dickinson, San Jose, CA). RBCs were also triply stained with PE-conjugated anti-CD55, anti-CD59 MoAb, and FITC-conjugated anti-glycophorin A MoAb (Immunotech) and analyzed by FACScan. Granulocytes and erythrocytes were identified on the basis of side scatter and forward scatter properties and positive staining with lineage-specific CD13 or glycophorin A antigens. CD55 and CD59 double negative cells in lineage specific antigen positive cells were calculated.

Results

The samples from healthy adults consistently had less than 0.01% GPI-AP–deficient cells in both the granulocyte and erythrocyte populations (0.001±0.001 and 0.002±0.003%). The equivalent figures for pre-IST patients were 0.02±0.041 and 0.194±0.469%, and in post-IST patients, 0.005±0.012 and 0.001±0.001%, respectively. In contrast, the patient with symptomatic AA-PNH syndrome had 18.8% of granulocytes and 57.8% of RBCs with GPI-AP deficiency. Only one patient in the pre-IST group had greater than 1% (1.5%) GPI-AP–deficient RBCs. More than 0.01% of GPI-AP–deficient granulocytes were detected in 3/10 and 3/20 of pre-and post-IST patients, respectively. More than 0.01% of GPI-AP–deficient RBCs were detected in 6/10 and 8/20 of pre-and post-IST patients, respectively.

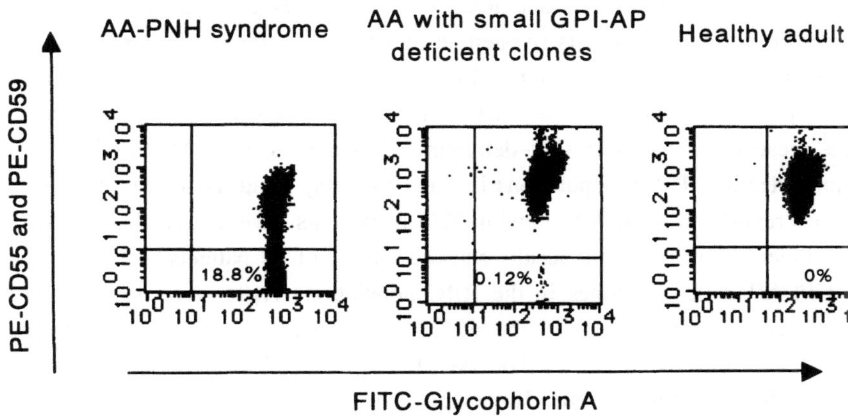

Figure.1. RBCs were triply stained with PE-conjugated anti-CD55, anti-CD59 MoAb, and FITC-conjugated anti-glycophorin A MoAb and analyzed by FACScan. GPI deficient cells are demonstrated in the lower right quadrant.

Discussion

The recently developed technique using flow cytometry to detect the GPI-AP–deficient clones in AA has brought a new understanding to the pathological relationship between PNH and AA. However, not much is known about the incidence and the clinical significance of GPI-AP–deficient blood cells in children with AA. We previously studied the deficiencies of GPI-AP on granulocytes by single staining of CD55 or CD59 in 116 children recently diagnosed (from 11/1996 to 10/2000) with AA (unpublished data). A significant GPI-A– deficient population (≧3% of both CD55 negative and CD59 negative granulocytes) was demonstrated in only 2 patients and no clinical PNH has occurred in these patients to date. To elucidate the relationship between presence of PNH clones and clinical features, we selected 12 AA patients with more than 2.5% of CD55 negative and 1.0% of CD59 negative cells, and compared their clinical characteristics and the response rate (RR) to IST with other patients. We found no significant differences regarding age, sex, and severity of AA, and RR between 2 groups (RR at 6 months; 54.5% vs. 58.2%). These results showed the frequency of GPI-AP–deficient blood cells is lower in AA children than in adult patients. This is consistent with the clinical finding that few AA children develop PNH after IST. Recently, with the use of more sensitive flow cytometry techniques, the presence of a small number of GPI-AP–deficient cells with PIG-A gene abnormalities were reported in normal individuals (Araten, et al 1999). We, therefore, conducted this second study using triple staining flow cytometric analysis, which can detect smaller population of PNH clones than our first study was able to do. The results of the present study showed GPI–deficient cells were less than 0.01% of both the granulocytes and RBCs populations in the 10 healthy controls. In contrast to the previous reports on adult AA showing 22% of patients have greater than 1% GPI-AP deficient clones [2], our results showed only 1 of 30 patients had more than 1% GPI-AP–deficient clones in the RBC population. There were no significant differences between patients in the pre-IST and post-IST groups. These results showed that the frequency of GPI-AP–deficient blood cells is lower in AA children than in adult patients. However, in our study, a small number of GPI-AP–deficient clones (> 0.01%) were demonstrated in some patients (12/30 patients in RBCs and 6/30 in granulocytes). Further studies will be necessary to elucidate the clinical significance of these populations in children with AA.

References

1. Schrezenmeier H, Hertenstein B, Wagner B, et al (1995) A pathogenetic link between aplastic anemia and paroxysmal nocturnal hemoglobinuria is suggested by a high frequency of aplastic anemia patients with a deficiency of phosphatidylinositol glycan anchored proteins. Exp Hematol 23:81-7

2. Dunn DE, Tanawattanacharoen P, Boccuni P, et al (1999) Paroxysmal nocturnal hemoglobinuria cells in patients with bone marrow failure syndromes. Ann Intern Med 131:401-8

3.De Lord C, Tooze JA, Saso R, et al (1998) Deficiency of glycosylphosphatidyl inositol-anchored proteins in patients with aplastic anaemia does not affect response to immunosuppressive therapy. Br J Haematol 101:90-3

4.Araten DJ, Nafa K, Pakdeesuwan K, et al (1999) Clonal populations of hematopoietic cells with paroxysmal nocturnal hemoglobinuria genotype and phenotype are present in normal individuals. Proc Natl Acad Sci USA 96:5209-14

Helper T Cell Subsets and Th1/2 Balance in Aplastic Anemia

Kensuke Usuki , Akiko Furuya , Yuji Hirai , Seiko Iki , and Akio Urabe

Division of Hematology, Kanto Medical Center, 5-9-22, Higashi-Gotanda, Shinagawa-ku, Tokyo 141-8625, Japan

Summary: Autoimmune pathophysiology of aplastic anemia has been implicated, since immunosuppressive therapy appeared to be effective. On autoimmune diseases, disturbance of Th1/2 balance has been suggested to play the critical role. We analyzed helper T cell subsets in 10 patients with aplastic anemia (6 patients of severe grade, 3 moderate, and 1 mild) using flowcytometer with antibodies against IFN-γ and IL-4 after permeabilization in the stimulated condition by TPA and Ionomycin in the presence of Brefeldin-A. The percentage of helper T cells elevated in 2 patients and reduced in 4 patients. The proportion of activated helper T cells elevated in 3 patients. The CD4/8 ratio decreased in 4 patients and increased in 1 patient. The percentage of Th1 cells increased in 3 patients. In one of these 3 patients, the Th1/2 ratio elevated. Another patient showed also the increased ratio of Th1/2. Thus, an elevation of the Th1 cell percentage or the Th1/2 ratio was detected in 4 out of 10 patients. On the contrary, the increased percentage of Th2 cells was seen only in 1 patient, in whom the Th1 cell proportion also elevated resulting in normal Th1/2 ratio. These results suggest that Th1 cells play an important role in phathogenesis of AA.

Key words: Aplastic anemia, Helper T cell, T cell subset, Th1/2 balance, Th1

Introduction

Autoimmune pathophysiology of aplastic anemia (AA) has been implicated, since immunosuppressive therapy appeared to be effective [1]. Helper T cells are divided into 3 subgroups, Th0, Th1, and Th2. On autoimmune diseases, disturbance of Th1/2 balance has been suggested to play the critical role [2]. We analyzed the helper T cell subsets in 10 patients with AA including 1 post-hepatitis AA.

Patients and Methods

Seven patients were male and 3 were female as shown in Table. The average age was 44.2 year-old (18~75). Six patients were newly diagnosed untreated, and the remaining 4 were in relapse or refractory to immunosuppressive therapy. Six patients were in severe grade, 3 were in moderate, and 1 was in mild.

T cell subsets in peripheral blood were analyzed using flowcytometer with monoclonal antibodies against CD3, 4, 8, and 20, and HLA-DR. For sub-classification of helper T cells, $CD4^+$ cells were permeabilized in the stimulated condition by TPA and Ionomycin in the presence of Brefeldin-A, and then analyzed with flowcytometer using antibodies against IFN-γ and IL-4.

Table. Patient characteristics

Ptn.	age	gender	diagnosis	grade	WBC ($/\mu l$)	neutrophils ($/\mu l$)	Hb (g/dl)	reticulocytes ($x10^4/\mu l$)	platelets ($x10^4/\mu l$)
1	28	M	N	severe post-hapatitis	4000*	3600*	9.3	5.7	3.4
2	74	F	N	severe	1400	105	8.3	1.1	0.7
3	22	F	N	severe	1600	80	8.4	0.3	0.7
4	20	M	N	severe	900	243	7.1	0.5	1.5
5	75	M	N	moderate	2500	1425	7.2	5.1	4.4
6	44	M	N	moderate	4700	1998	5.8	4.4	3.1
7	53	M	R	severe	800	304	5.6	0.2	2.2
8	18	M	R	severe	5800	4930	7.3	1.7	0.7
9	37	M	R	moderate	3100	2000	6.0	3.2	2.4
10	71	F	R	mild	2300	1311	11.4	7.7	2.3

* receiving G-CSF; Ptn., patient number; N, newly diagnosed previously untreated; R, relapse or refractory to immunosuppressive therapy.

Results

As shown in Fig., the percentage of T cells ($CD3^+$ cells) elevated in 4 patients including 3 previously untreated patients (Ptn. 2, 3, and 6) and 1 relapse patient (Ptn.10). The average of the T cell proportions in total 10 patients with AA (mean $\pm$SD; 88.6$\pm$5.2%) slightly increased (normal range; 66~89%). The percentage of B cells ($CD20^+$ cells) decreased slightly in 2 previously untreated patients (Ptn. 2 and 3) and increased slightly in 3 patients including 1 previously untreated patient (Ptn. 4) and 2 relapse patients (Ptn. 7 and 8). The average of B cell proportions in 10 patients (9.0$\pm$5.1%) was within normal range (4~13%). The percentage of helper T cells ($CD4^+$ cells) increased in 2 previously untreated patient (Ptn. 2 and 3) and decreased in 4 patients including 2 previously untreated patients (Ptn. 1 and 5) and 2 relapse patients (Ptn. 8 and 9). The average of helper T cell proportions in

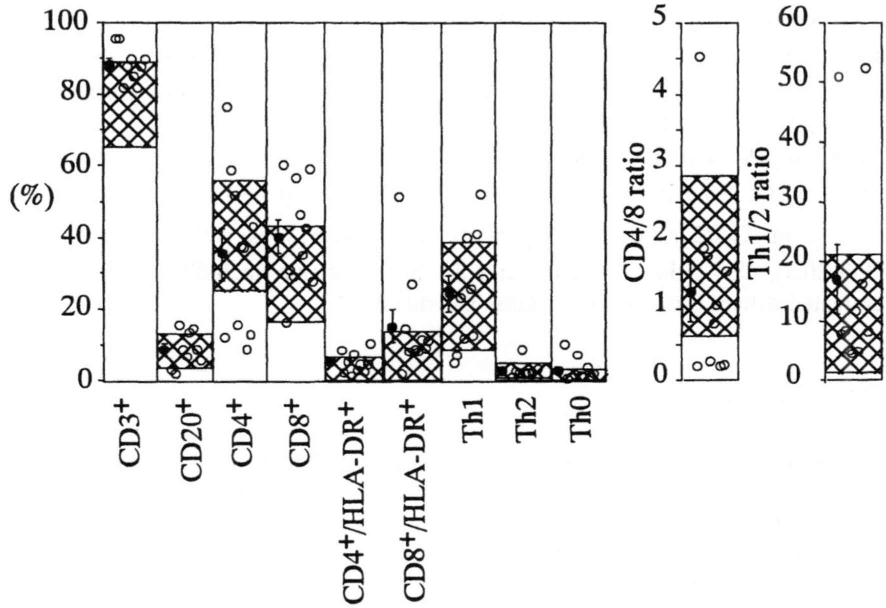

Fig. Lymphocyte subsets in aplastic anemia. Open circles indicate values of each patient. A close circle with bars indicates mean ± standard deviation. Hatched area indicates normal range.

10 patients (35.7±22.8%) was within normal range (25~56%). The percentage of activated helper T cells ($CD4^+$/HLA-DR^+ cells) increased slightly in 3 patients including 2 previously untreated patients (Ptn. 1 and 5) and 1 relapse patient (Ptn. 10). Their average in 10 patients (5.2±3.1%) was within normal range (≦7%). The suppressor T cell ($CD8^+$ cells) percentage decreased slightly in 1 previously untreated patient (Ptn. 2) and increased in 4 patients including 3 previously untreated patients (patients 1, 5, and 6) and 1 relapse patient (Ptn. 9). Their average in 10 patients (40.7±15.0%) was within normal range (17~44%). Three previously untreated patients (Ptn. 1, 3, and 5) showed increase of the activated suppressor T cell ($CD8^+$/HLA-DR^+ cells) percentage. Its average in 10 patients (15.4±14.4%) slightly increased (normal range: <14%). Although the CD4/8 ratio in 10 patients (1.25±1.33) was within normal range (0.6~2.9), the ratio increased in 1 previously untreated patient (Ptn. 2) and decreased in 4 patients including 2 previously untreated patients (Ptn. 1 and 5) and 2 relapse patients (Ptn. 8 and 9). The Th1 cell (INF-γ^+/IL-4^-cells) percentage increased in 3 patients including 1 previously untreated patient (Ptn. 5) and 2 relapse patients (Ptn. 8 and 9), and

reduced slightly in 2 previously untreated patients (Ptn. 1 and 2). The average percentage of Th1 cells in 10 patients (25.0±16.0%) was within normal range (8.9~39.1%). The percentage of the Th2 cells (INF-γ^-/IL-4^+ cells) increased in 1 previously untreated patient (Ptn. 5) and decreased slightly in 1 previously untreated patient (Ptn.1). The average of the Th2 cell proportions in 10 patients (mean±SD: 2.7±2.4%) was within normal range (0.6~4.3%). Although the average of Th1/2 ratios in 10 patients (17.0±18.6) was within normal range (1.4~20.5), the ratio elevated markedly in 2 patients including 1 previously untreated patient (Ptn. 1) and 1 relapse patient (Ptn. 9).

Discussion

The percentage of Th1 cells increased in 3 patients (Ptn. 5, 8, and 9). In one (Ptn. 9) of these 3 patients, the Th1/2 ratio elevated. Another patient (Ptn. 1) showed also the increased Th1/2 ratio. On the contrary, the increased percentage of Th2 cells was seen only in one patient (Ptn. 5), in whom the Th1 cell proportion also elevated resulting in normal Th1/2 ratio. Thus, an elevation of the Th1 cell percentage or the Th1/2 ratio was detected in 4 out of 10 AA patients. These results suggest that Th1 cells play an important role in phathophysiology of AA, consistent with the fact that Th1 cells contribute to the pathogenesis of organ-specific autoimmune disease [2]. On the other hand, half of patients represented no disturbance of Th1/2 balance. Maciejewski et al reported that in AA the increased percentages of activated $CD8^+$ cells were found more frequently in marrow than in peripheral blood [3]. Furthermore, the marked discrepancy between expression of IFN-γ in peripheral blood and marrow from AA patients implies the restriction of the pathologic process to the target organ [4]. It remains to explore the possibility that Th1/2 balance is disturbed more frequently in bone marrow which is target organ of autoimmunity in aplastic anemia.

References

1. Young NS (2000) Hematopoietic cell destruction by immune mechanisms in acquired aplastic anemia. Semin Hematol 37: 3-14
2. Liblau RS, Singer SM, McDevit HO (1995) Th1 and Th2 $CD4^+$ T cells in the pathogenesis of organ-specific autoimmune diseases. Immunol Today 16: 34-38
3. Maciejewski JP, Hibbs JR, Anderson S, et al (1994) Bone marrow and peripheral blood lymphocyte phenotype in patients with bone marrow failure. Exp Hematol 22: 1102-1110
4. Nakao, S, Yamaguchi M, Shiobara S, et al (1992) Interferon-γ gene expression in unstimulated bone marrow mononuclear cells predicts a response to cyclosporine therapy in aplastic anemia. Blood 79: 2532-2535

Relationship between immunologic abnormalities and paroxysmal nocturnal hemoglobinuria-associated clones in myelodysplastic syndromes

Takahiro Okamoto, Masaya Okada, Ako Mori, Hiroyuki Takatsuka, Yoshihiro Fujimori and Eizo Kakishita

Division of Hematology and Oncology, Department of Internal Medicine, Hyogo College of Medicine

Summary: Blood cells representing a paroxysmal nocturnal hemoglobinuria (PNH)-associated clone have been reported in patients with myelodysplastic syndromes (MDS). Since immunologic abnormalities are frequent in MDS, we investigated relationships between immune dysfunction and PNH clones. Erythrocyte and granulocyte fractions from 38 MDS patients were examined by flow cytometry to detect PNH clone cells, which lack CD55 and CD59 antigens. Six patients, all among is showing immunologic abnormaliteis, had PNH clones constituting 1 to 5% of blood cells; one also had rheumatoid arthritis. As a significant association was found between PNH clones and immunologic abnormalities (p=.0009), PNH clones may reflect immune disturbance in MDS.

Keywords: paroxysmal nocturnal hemoglobinuria (PNH) clone, immunologic abnormality, myelodysplastic syndromes

Introduction

An extremely small fraction of blood cells representing a paroxysmal nocturnal hemoglobinuria (PNH)-associated clone has been demonstrated in eight normal individuals [1]. However, a PNH clone does not expand to greater numbers in the normal state. Aplastic anemia is considered to have a mainly autoimmune pathogenesis, since immunosuppressive treatment can permit recovery of hematopoiesis [2,3]. More than 50% of patients with aplastic anemia are believed to develop a sizable PNH clone, suggesting that growth of the PNH clone is related to disturbed immune function [4]. Such PNH clones have been found in a small percentage of patients with myelodysplastic syndrome (MDS) as well [5,6].

This paper first appeared in the British Journal of Haematology (2001) Sep;114(3):731-733. The present contents are unchanged, but the data have been updated and the figure has been revised.

MDS frequently is associated with immune abnormalities, such as abnormal results of immunologic tests or autoimmune disorders [7,8]. We therefore investigated the relationship between existence of a PNH clone and immunologic abnormalities.

Patients and methods

Patients

Thirty-eight patients, all fulfilling diagnostic criteria for MDS established by the French-American-British cooperative study group [9], were examined in this study after giving informed consent. Patient characteristics are described in the Table. Ages of patients ranged from 18 to 84 years (median, 59). Among 38 patients, 33 had refractory anemia (RA), 1 had refractory anemia with ringed sideroblasts (RARS), and 4 had refractory anemia with excess of blasts (RAEB). Immunologic abnormality was defined as any of the followings: antinuclear antibody, rheumatoid factor, anti-DNA antibody, positive Coombs test, elevation of immunoglobulin, or a CD4/CD8 ratio among T lymphocytes of less than 1.

Cell preparation

Peripheral blood samples from MDS patients were collected in heparinized tubes. The blood was layered onto Ficoll-Paque (ρ =1.007; Pharmacia Biotech, Sweden), and centrifuged at 400 x g for 20 min. Erythrocytes were collected in the bottom layer. The upper most layer of red cells, containing granulocytes, was collected and the suspended in Tris-NH_4Cl buffer to lyse erythrocytes. The purity of granulocytes exceeded 98% according to light microscopic examination after May-Grunwald-Giemsa staining. These cell separation procedures were performed to increase purity of each cell population for two-color flow cytometric analysis.

Antibody staining and flow cytometric analysis

Phycoerythrin (PE)-conjugated mouse anti-human CD55 monoclonal antibody (MoAb; IgG1) was purchased from Immunotech (Marseille, France). Fluorescein isothiocyanate (FITC)-conjugated mouse anti-human CD59 MoAb (IgG1) was obtained from Serotec (Oxford, England). The fraction of erythrocytes or granulocytes was incubated on ice with PE-CD55 MoAb for 30 min, washed twice with phosphate-buffered saline (PBS), and then incubated on ice with FITC-CD59 MoAb for 30 min. The cells were washed twice in PBS and then analyzed by flow cytometry using a EPICS-XL system II analyzer (Beckman Coulter, Inc, Fullerton, CA). Forward and side scatter gating for erythrocytes or

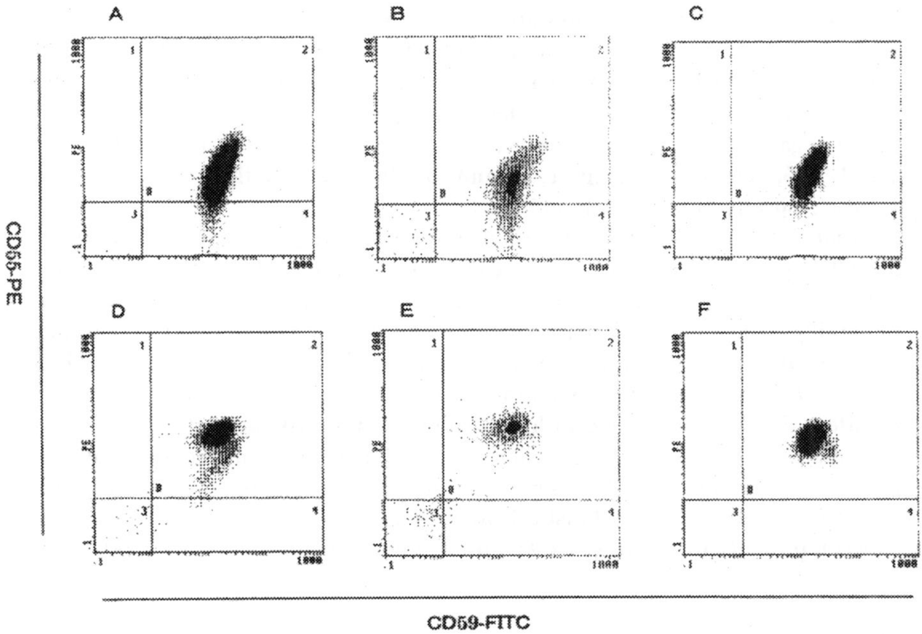

Figure Two-color flow cytometry using CD55-PE and CD59-FITC in erythrocytes and granulocytes. A, erythrocytes and D, granulocytes in patient 1; B, erythrocytes and E, granulocytes in patient 4; C, erythrocytes and F, granulocytes in a normal volunteer.

granulocytes was strictly selected. Nonspecific adsorption of antibodies to Fc receptors on examined cells was blocked by preincubation of cells with human serum from healthy volunteers having AB blood type. PE- and FITC-conjugated mock antibody (IgG1) was used for setting the gate to exclude negative cells.

Statistics

The χ^2 test was used to examine the association between existence of a PNH clone and immunologic abnormalities.

Results

As shown in the Figure, two-color flow-cytometric analysis for CD55 and CD59 indicated very few or no cells negative for both CD55 and CD59 in a normal individual (panels C and F). The CD55-negative, CD59-negative populations of erythrocytes from 10 normal volunteers ranged from 0.00 to 0.05%, with a median of 0.02%; those of granulocytes ranged from 0.00 to 0.05%, with a median of 0.00%. Thus, we determined that a CD55-negative, CD59-negative population exceeding 0.1% indicated existence of a PNH clone. As for representative patients, two-color analysis indicated CD55-negative, CD59-negative populations of erythrocytes and granulocytes in patient 1 as 1.23 and 1.93%, respectively (Figure, A and D); in patient 4 these were 4.81 and 1.46%, respectively (Figure, B and E).

Results of flow cytometric analysis in all patients are presented in the Table. Of 38 MDS patients, 6 (16%) had a CD55-negative, CD59-negative population, representing a PHN clone in both erythrocytes and granulocytes; and the remaining 32 had no PNH clone evident in either cell fraction. On the other hand, 14 of 38 patients had an abnormal immunologic test result, and all 6 patients with PNH clone fell into this group. One of these (patient 2) had clinical manifestations of rheumatoid arthritis. None of the 6 patients with a PNH clone had recently received a transfusion of red blood cells at the time of flow cytometric examination, although a few of the patients showing no clone had recently been transfused. Statistical analysis indicated a significant association between presence of a PNH clone and immunologic abnormalities (p=.0009).

Discussion

Patients with MDS are more likely to have abnormal results of immunologic tests and occurrence of connective tissue disease than the normal population [7]. However, MDS shows a PNH clone in only a small fraction of cases compared to aplastic anemia, with more than 50% of patients showing a clone [4]. In the present study, we found that appearance of a PNH clone in MDS is associated with abnormal immunologic laboratory findings.

Iwanaga et al [5] reported that flow cytometry showed CD59-negative erythrocytes and granulocytes in 4 of 40 patients with MDS (10%); with proportions of CD59-negative erythrocytes and granulocytes being 12 to 22% (median, 20) and 16 to 95% (median, 17). In our study, 6 of 38 patients (16%) had CD55-negative, CD59-negative cells. Proportions of CD55-negative, CD59-negative erythrocytes and granulocytes respectively were 0.10 to 4.81% (median, 0.71) and 0.10 to 1.94% (median, 0.64). The proportion of CD59-negative cells reported in the previous investigation was higher than that of CD55-negative, CD59-negative cells in our study. We believe that the proportion of negative cells found by Iwanaga et al [5] would have been lower if two-color analysis of CD55 and CD59 MoAb had been performed.

Disease types in our patients having a PNH clone were either RA or RARS; none of four patients with RAEB had a PNH clone. In Iwanagaís results, only patients with RA had a PNH clone, while those with RAEB, RAEB in transformation or chronic myelomonocytic leukemia did not. PNH clones, then, might occur only in low-risk MDS without increased myeloblasts. Larger

Table. Patient characteristics and the finding of paroxysmal nocturnal haemoglobinuria-associated clone myelodysplastic syndromes

Pt	Age/gender	Disease class[1]	Karyotype	Treatment[2]	Time after diagnosis (months)	Bone marrow cellularity[3]	Immunologic abnormalities[4]	CD55 & CD59-negative population (%)	
								erythrocytes	granulocytes
1	45/M	RARS	46,XY,del(20)(q11;q13.3) [19/20]	AS, PSL	5	normo	+ : direct Coombs test (+), IgE 1331 mg/dL	1.23	1.94
2	64/F	RA	normal	AS, PSL	83	hypo	+ : RA, RA factor 31.6IU/mL	0.11	0.44
3	69/M	RA	46,XY,+8, -Y [16/20]	Cy A, AS	26	normo	+ : ANA x80, IgG 2010 mg/dL	0.10	0.80
4	57/M	RA	normal	AS	78	normo	+ : Anti-parietal cell ab (+), ANA x40	4.81	1.46
5	63/F	RA	normal	AS	55	normo	+ : ANA x320	2.83	0.48
6	60/F	RA	46,XX,13q- [20/20]	AS	48	hypo	+ : ANA x120	0.18	0.10
7	69/F	RA	complex[5]	AS, Cy A	39	normo	+ : ANA x160, elevation of IgG, IgA	0.00	0.00
8	45/M	RA	45,X,-Y [3/20]	Cy A, AS	23	normo	+ : ANA x80	0.06	0.02
9	53/F	RA	46,XX,t(1:16)(p32:p11) [9/20]	AS	48	hypo	+ : ANA x80, CDT4/CDT8 <1	0.00	0.01
10	23/M	RA	46,XY,inv9(p11; q13)[6] [20/20]	AS, Cy A	144	hypo	+ : CD4/CD8 <1	0.00	0.00
11	79/F	RA	normal	AS	125	hypo	+ : anti-DNA ab x160	0.00	0.04
12	59/M	RA	normal	AS, PSL	89	hypo	+ : ANA x80	0.02	0.00
13	33/M	RA	normal	AS, G-CSF, Cy A	156	hypo	+ : CD4/CD8 <1, decreased IgG, IgA	0.03	0.05
14	53/M	RA	normal	AS, PSL	112	normo	+ : ANA x80	0.02	0.02
15	67/M	RA	normal	AS, Epo, G-CSF	88	hypo	+ : CD4/CD8 <1	0.02	0.02
16	72/M	RA	46,XY,del(8)(p11:p21)[12/20]	AS, Cy A	96	normo	-	0.02	0.02
17	55/M	RA	46,XY,del(20)(q11:q13) [7/20]	Cy A, AS, PSL	420	normo	-	0.00	0.00
18	72/M	RAEB	47,XY,+8,del(20)(q11:q13.3) [2/20]	no therapy	16	hypo	-	0.01	0.02
19	45/F	RAEB	not detected[7]	no therapy	9	hypo	-	0.03	0.01
20	76/M	RAEB	not detected	no therapy	9	hypo	-	0.00	0.00
21	73/M	RAEB	not detected	AS	20	hypo	-	0.06	n.d.[8]
22	79/M	RA	45,X,-Y [4/20]	AS	87	normo	-	0.00	0.01
23	66/M	RA	not detected	Cy A, AS, PSL	205	hypo	-	0.00	0.00
24	82/F	RA	46,XX,del(5q) [6/20]	AS, PSL	55	hypo	-	0.02	0.00
25	59/F	RA	normal	AS, Cy A, G-CSF, PSL	84	normo	-	0.02	0.04
26	53/M	RA	normal	no therapy	33	normo	-	0.04	0.02
27	18/M	RA	normal	G-CSF, AS, Cy A	37	normo	-	0.00	0.00
28	56/F	RA	normal	AS, Cy A, G-CSF, Epo	204	hypo	-	0.00	0.00
29	21/F	RA	normal	AS	153	hypo	-	0.00	0.02
30	70/M	RA	normal	AS	6	hyper	-	0.02	0.00
31	49/M	RA	normal	AS, PSL	106	hypo	-	0.03	0.00
32	22/M	RA	normal	AS	10	normo	-	0.00	0.04
33	65/M	RA	normal	AS	97	hypo	-	0.00	0.00
34	84/M	RA	normal	AS	25	hypo	-	n.d.	0.04
35	49/F	RA	normal	AS, PSL	18	normo	-	0.02	0.00
36	34/F	RA	normal	AS	112	normo	-	0.00	0.00
37	72/M	RA	normal	AS, PSL	55	hypo	-	0.02	0.02
38	27/F	RA	normal	AS	92	normo	-	0.00	0.00

1) Class: RA, refractory anaemia; RARS, refractory anaemia with ringed sideroblasts; RAEB, refractory anaemia with excess of blasts
2) Treatment: none, no therapy; AS, anaboric steroid; Cy A, cyclosporin A; PSL, prednisolone; Epo, erythropoietin
3) hypo, hypoplasia; normo, normoplasia; hyper, hyperplasia.
4) Immunologic abnormalities: RA, rheumatoid arthritis; ANA, anti-nucleic antibody; anti-DNA ab, anti-DNA antibody
5) 46,XX,del(2),del(5),del(11)[5/20]/46,idem,-8,+mar1[2/20]/47,idem,-8,+mar1,+mar2[1/20]/46,XX[12/20]
6) inv9(p11; q13) is thought to be a normal variant
7) not detected; mitotic chromosomes could not acquired for karyotype analysis.
8) n.d.: not done

numbers of MDS patients need to be analyzed to confirm this impression.

Dunn et al [6] reported that 9 of 39 (23%) patients with myelodysplasia had granulocytes deficient in GPI-anchored protein (CD16-negative, CD66b-negative populations). Patients with myelodysplasia having GPI-deficient granulocytes significantly showed a better response to treatment with antithymocyte globulin. In the present study, several patients had been treated with cyclosporine A (Cy A) as an immunosuppressant (Table); patient 3 who had PNH clone cells, responded well to Cy A treatment. However, meaningful correlation of PNH clones with clinical response to Cy A is not feasible in a small number of patients.

Recently the presence of extremely low numbers of GPI-deficient neutrophils in normal individuals has been described [1]. Whether the growth advantage of GPI-deficient cells in PNH is intrinsic to PNH stem cells or instead results from escape from an extrinsic regulatory process has been investigated. Study of a murine model suggested that a PIG-A mutation alone could not account for increased numbers of mutant stem cells [10]. One reason for expansion of a PNH clone might be escape from an immune-mediated suppressive mechanism. The association of a PNH clone in MDS patients with immunologic abnormalities in our study suggests that a relative growth advantage for the PNH clone might be conferred by immune dysfunction.

Acknowledgments

The authors are grateful to Mrs. Aki Nomura for invaluable assistance with cell preparations and flow cytometric analysis.

References

1. Araten DJ, Nafa K, Pakdeesuwan K, et al (1999) Clonal populations of hematopoietic cells with paroxysmal nocturnal hemoglobinuria genotype and phenotype are present in normal individuals. Proc Natl Acad Sci USA 96: 5209-5214
2. Bacigalupo A, Broccia G, Corda G, et al (1995) Antilymphocyte globulin, cyclosporin, and granulocyte colony-stimulating factor in patients with acquired severe aplastic anemia (SAA): A pilot study of the EBMT SAA Working Party. Blood 85: 1348-1353
3. Ball SE (2000) The modern management of severe aplastic anaemia. Br J of Haematol 110: 41-53
4. Schrezenmeier H, Hertenstein B, Wagner B, et al (1995) A pathogenetic link between aplastic anemia and paroxysmal nocturnal hemoglobinuria is suggested by a high frequency of aplastic anemia patients with a deficiency of phosphatidylinositol glycan anchored proteins. Exp Hematol 23: 81-87
5. Iwanaga M, Furukawa K, Amenomori T, et al (1998) Paroxysmal nocturnal haemoglobinuria clones in patients with myelodysplastic syndromes. Br J of Haematol 102: 465-474

6. Dunn DE, Tanawattanacharoen P, Boccuni P, et al (1999) Paroxysmal nocturnal hemoglobinuria cells in patients with bone marrow failure syndromes. Ann Intern Med 131: 401-408
7. Hamblin T (1992) Immunologic abnormalities in myelodysplastic syndromes. Hematol Oncol Clin North Am 6: 571-586
8. Okamoto T, Okada M, Mori A, et al (1997) Correlation between immmunological abnormalities and prognosis in myelodysplastic syndrome patients. Int J Hematol 66: 345-352
9. Bennett JM, Catovsky D, Daniel MT, et al (1982) Proposals for the classification of the myelodysplastic syndromes. Br J of Haematol 51: 189-199
10. Murakami Y, Kinoshita T, Maeda Y, et al (1999) Different roles of glycosylphosphatidylinositol in various hematopoietic cells as revealed by a mouse model of paroxysmal nocturnal hemoglobinuria. Blood 94: 2963-2970

Repeated Response to Antithymocyte Globulin in a Case of Aplastic Anemia-Paroxysmal Nocturnal Hemoglobinuria Syndrome

Takeshi Yasunami , Masanao Teramura , Koji Iwabe , Michihiko Masuda , Toshiko Motoji , Hideaki Mizoguchi

Department of Hematology, Tokyo Women's Medical University, 8-1 Kawada-cho, Sinjuku-ku, Tokyo 162-8666, Japan

Summary: We report a patient with aplastic anemia-paroxysmal nocturnal hemoglobinuria (AA-PNH) syndrome who was successfully treated with repeated administration of antithymocyte globulin (ATG). The patient became transfusion-independent after a first course of horse ATG, however, became transfusion-dependent again after nine month. The patient received a second course of horse ATG, resulting in the improvement of anemia and thrombocytopenia again. The patient is currently transfusion-independent with mild pancytopenia and PNH. This case clearly shows that repeated treatment with ATG is a good strategy even in the recurrent cases of AA-PNH syndrome.

Key words: Antithymocyte globulin, Aplastic Anemia-Paroxysmal Nocturnal Hemoglobinuria Syndrome, repeated treatment

Introduction

Paroxysmal nocturnal hemoglobinuria (PNH) is an acquired clonal hematological stem cell disorder characterized by complement-mediated hemolysis and the expansion of affected cells of various hematopoietic lineages. It is sometimes associated with aplasitic anemia (AA) and called AA-PNH syndrome. There are 8-32% of AA patients complicating PNH [1-3]. Therefore AA-PNH syndrome may not be a rare disorder. As for AA, efficiency of immunosuppressive agents including antithymocyte globulin (ATG), antilymphocyte globulin (ALG) and cyclosporine have been established, however, few reports concerning the immunosuppressive therapy for PNH or AA-PNH syndrome have been reported [4-8]. Repeated treatment with ATG/ALG has also proved to be useful for AA [9,10], but not for PNH or AA-PNH syndrome. In this report, we present a case of AA-PNH syndrome successfully treated with repeated administration of ATG.

Case report

A 55-year-old woman was admitted to another hospital because of fatigue in April

1998. She had pancytopenia (white blood cell 2 x 10^9/L, hemoglobin 5.8g/dl, platelet 11 x 10^9/L) and aspirated bone marrow showed hypoplasia. A diagnosis of AA was made, and she was treated with cyclosporine A (6mg/kg) without response. She needed red blood cell transfusion once or twice a month. In Feb 1999, she was admitted to our hospital. Her white blood cell count was 1.9 x 10^9/L, hemoglobin 7.5g/dl, platelet 14 x 10^9/L. Serum total bilirubin level was 0.8mg/dl, lactate dehydrogenase 544IU/l and haptoglobin <10mg/dl. Sucrose hemolysis and Ham test were positive, and flow cytometric study showed diminished expression of CD59 on red blood cell (88.1%). The diagnosis of AA-PNH syndrome was made. She was treated with horse ATG (15mg/kg for 5days) and became transfusion-independent one month after ATG treatment. But 9 months later, her hemoglobin and platelet counts gradually decreased again and began to require regular red blood cell transfusion. Bone marrow aspiration showed hypercellularity with erythroid predominance. Increased serum lactic dehydrogenase level (701IU/l), and CD55- and CD59- negative red cell population (21.2% and 17.9% respectively) suggested the expansion of PNH clone. She was treated with a second course of horse ATG treatment (15mg/kg for 5days) in Feb 2000. She had no severe adverse effect during the ATG treatment, although mild serum sickness occurred. Within 1 month after treatment, she became transfusion-independent again.

The patient is currently transfusion-independent with mild pancytopenia (white blood cell 2 x 10^9/L, hemoglobin 9.8g/dl, platelet 53 x 10^9/L) and persistent presence of PNH.

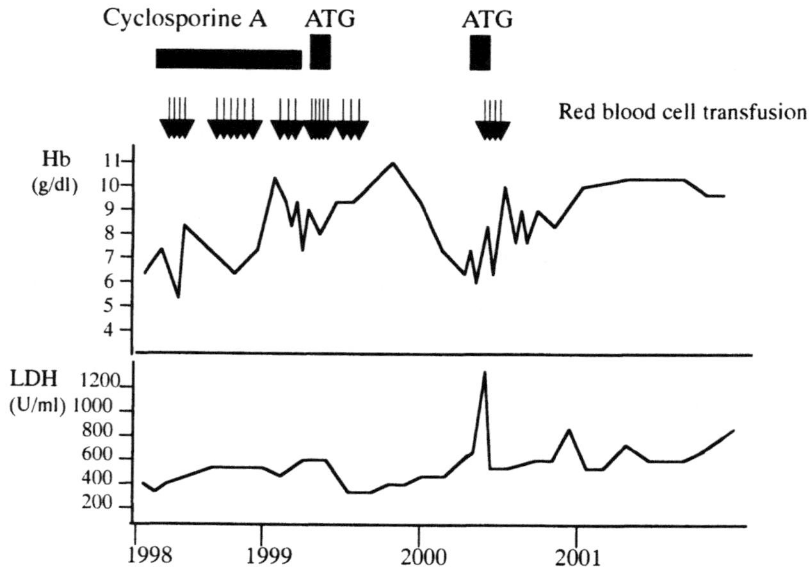

Fig. 1. clinical course of the patient

Discussion

Several immunosuppressive agents including ATG and cyclosporine are used for the treatment of patient with PNH [4,5]. Paquette et al. have reported that three out of seven patients with PNH treated with ATG experienced a sustained improvement in at least one peripheral blood cytopenia [4]. The authors emphasize that all responding patients had hypoplastic bone marrow and a minor degree of chronic hemolysis. Others reported that three patients with PNH evolving into severe AA responded to cyclosporine therapy [5]. As for AA-PNH syndrome, Kamp et al. reported two patients successfully treated with cyclosporine or ATG [6]. From these reports the patient who responded to immunosuppressive therapy might have PNH with hypoplastic bone marrow.

It has been shown that repeated treatment with ATG/ALG is effective for AA patients who were unresponsive to a first course of ATG/ALG or relapsed AA after a first course of ATG [9,10]. Tichelli et al. showed 43 cases of refractory or relapsed AA after first ALG treatment [9]. Patients received second or subsequent course of ALG. Transfusion-independent haemopoiesis was achieved in 63%. And the incidence of serum sickness was 63% after the initial course compared to 57% after subsequent courses. Di Bona et al. reported that 30 patients with severe AA not responding to a first course of horse-ALG, cyclosporine and granulocyte colony stimulating factor, received a second course using rabbit-ATG, cyclosporine and granulocyte colony stimulating factor [10]. Transfusion-independence was achieved 77% patients, and side effects were mild.

However it is not established whether repeated treatment of ATG/ALG is also effective for patients with AA-PNH syndrome. In this report, we presented a case of AA-PNH syndrome successfully treated with second course of ATG. Our observation suggests that repeated administration of ATG is tolerable and effective for the recurrent cases of AA-PNH syndrome.

References

1. Griscelli-Bennaceur A, Gluckman E, Scrobohaci ML, et al (1995) Aplastic anemia and paroxysmal nocturnal hemoglobinuria: search for a pathogenetic link. Blood 85:1354-63
2. Tichelli A, Gratwohl A, Nissen C, et al (1994) Late clonal complications in severe aplastic anemia. Leuk Lymphoma 12:167-75
3. Najean Y, Haguenauer O (1990) Long-term (5 to 20 years) Evolution of nongrafted aplastic anemias. The Cooperative Group for the Study of Aplastic Refractory Anemias. Blood 76:2222-28
4. Paquette RL, Yoshimura R, Veiseh C, et al (1997) Clinical characteristics predict response to antithymocyte globulin in paroxysmal nocturnal haemoglobinuria. Br J Haematol 96:92-97
5. Stoppa AM, Vey N, Sainty D, et al (1996) Correction of aplastic anaemia complicating

paroxysmal nocturnal haemoglobinuria: absence of eradication of the PNH clone and dependence of response on cyclosporin A administration. Br J Haematol 93:42-44

6. Kamp H, Imhoff GW, Wolf JTM, et al (1995) The effect of cyclosporine on haematological parameters in patients with paroxysmal nocturnal haemoglobinuria. Br J Haematol 89:79-82
7. Kusminsky GD, Barazzutti L, Korin JD, et al (1988) Letter to editor: Complete response to antilymphocyte globulin in a case of aplastic anemia-paroxysmal nocturnal hemoglobinuria syndrome. Am J Hematol 29:123
8. Ebenbichler CF, Wurzner R, Sandhofer AD, et al (1996-97) Anti-thymocyte globulin treatment of a patient for paroxysmal nocturnal haemoglobinuria-aplastic anaemia syndrome: complement activation and transient decrease of the PNH clone. Immunobiology 196:513-21
9. Tichelli A, Passweg J, Nissen C, et al (1998) Repeated treatment with horse antilymphocyte globulin for severe aplastic anaemia. Br J Haematol 100:393-400
10. Di Bona E, Rodeghiero F, Bruno B, et al (1999) Rabbit antithymocyte globulin (r-ATG) plus cyclosporine and granulocyte colony stimulating factor is an effective treatment for aplastic anaemia patients unresponsive to a first course of intensive immunosuppressive therapy. Gruppo Italiano Trapianto di Midollo Osseo (GITMO). Br J Haematol 107:330-334

Treatment of PNH hemolytic crisis with heparin or low-molecular weight heparin by its dual (anti-complement and anti-coagulant) activity

Chikashi YOSHIDA[1], Yasushi OHKOSHI[1], Yuichi HASEGAWA[1], Toshiro NAGASAWA[1], Haruhiko NINOMIYA[2]

[1]Division of Hematology, Institute of Clinical Medicine; [2]College of Medical Technology, University of Tsukuba, Tennodai 1-1-1, Tsukuba 305-8577, Japan

Summary: The anti-complement (C') activity of heparin has been well demonstrated, but its clinical use as anti-C' drug has been limited by its anticoagulant activity. A 34-year-old man with PNH was complicated by two episodes of hemolytic crisis in 1999 and in 2000, triggered by infections. Because hemolytic crisis is often followed by thrombotic complications in PNH, we treated the crises with heparin or low-molecular weight heparin (dalteparin) by expecting dual effects on the C' and coagulation systems. On the first crisis, heparin was administered at 8,000 IU/day x 9 days which obtained 0.16-0.28 IU/mL anti factor Xa activity in the plasma. On the second crisis, dalteparin was administered at 4,800 U/day x 20 days which obtained 0.04-0.07 U/mL anti-factor Xa activity. On the both crises, no thrombosis developed. Red cell transfusions were required under heparin administration, but not under dalteparin. In the both courses of crises, biochemical data returned to the levels before the crises in 14-20 days. In vitro experiment demonstrated that heparin or dalteparin (> 0.6 IU/mL) effectively inhibit C'-mediated hemolysis in PNH. Because anti-C' activity of heparin is independent of its anticoagulant (AT-III-binding) activity, heparin or its derivatives with enriched anti-C' activity are potentially useful to inhibit C'-mediated hemolysis as well as to prevent thrombotic complications in PNH.

Key words: heparin, low-molecular weight heparin, thrombosis, complement

Introduction

Heparin is widely used clinically as an anticoagulant [1]. Only about one-third of heparin preparations has affinity for antithrombin III (ATIII) and this fraction is responsible for the anticoagulant activity of heparin [2]. Heparin has multiple biological properties, one of which is to regulate complement (C') pathways in vitro [3] and in vivo [4]. The anticoagulant effect of heparin is mediated by its binding to ATIII and activation of ATIII against thrombin and factor Xa. The C'-mediated lysis of paroxysmal nocturnal hemoglobinuria (PNH) red blood cells (RBC) is due to the deficiency of glycosyl phosphatidylinositol-anchored proteins with C'-regulatory activity [5]. We reported the effects of heparin or low-molecular weight heparin (LMWH) on the inhibition of C'-mediated hemolysis in PNH in vitro [6]. Here, we discuss the potential usefulness of heparin and its derivatives in PNH from the experience of heparin- and LMWH-administration upon the two episodes of hemolytic crisis.

Case and Methods

Case of PNH

Thiry-four-year-old man with PNH was complicated by hemolytic crisis triggered by to viral infection in 1999 and in 2000.

Materials

Heparin (1,000 U/ml) was from Novo Nordisk Pharma (Copenhagen, Denmark); dalteparin (Fragmin, 1,000 IU/ml) was from Kissei Pharmaceutical (Matsumoto, Japan). Human ATIII (Aventis Pharma) coupled with CNBr-Sepharose 4B (Amersham Pharmacia) was prepared, then ATIII-affinity and –non-affinity fractions of heparin or its derivatives were separated using ATIII-affinity chromatography.

C' sensitivity assay

The anti-C' activity of heparin or its derivatives was examined according to the method described previously [6]. Activation of coagulation cascade under the C'-activation was assessed by the generation of prothrombin fragment F1+2 in plasma-based hemolytic assay in sucrose buffer containing 2 mM $CaCl_2$.

Activated partial thromboplastin time (APTT)

The effects of heparin or its derivatives on the APTT prolongation was examined by APTT of normal human plasma (Ci-Trol). The effects on APTT by fractions of heparin or its derivatives were compared on the basis of their uronic acid content, which was determined by the carbazole method of Bitter & Muir [7] with D-glucuronolactone as the standard.

Results

Clinical course

On the first crisis, heparin was administered at 8,000 IU/day x 9 days which obtained 0.16-0.28 IU/mL anti-factor Xa activity in the plasma. On the second crisis, dalteparin was administered at 4,800 U/day x 20 days which obtained 0.04-0.07 U/mL anti-factor Xa activity.

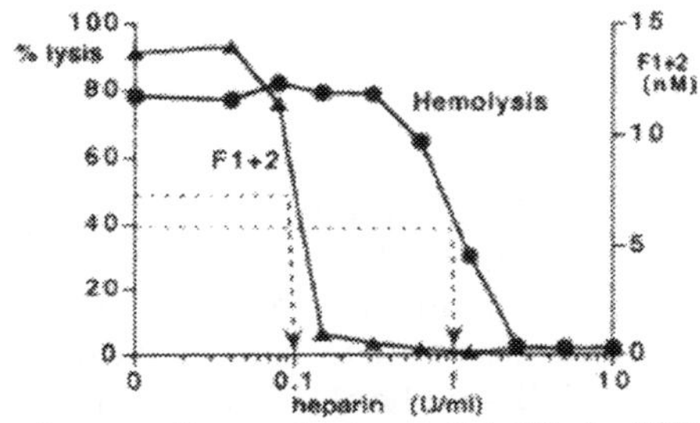

Fig. 1 IC50 of heparin as anticoagulant was 0.1 U/mL; IC50 of heparin as anti-C' was 1.0 U/ml. This experiment was made by plasma-based hemolytic assay of PNH RBC in sucrose buffer containing 2 mM CaCl2.

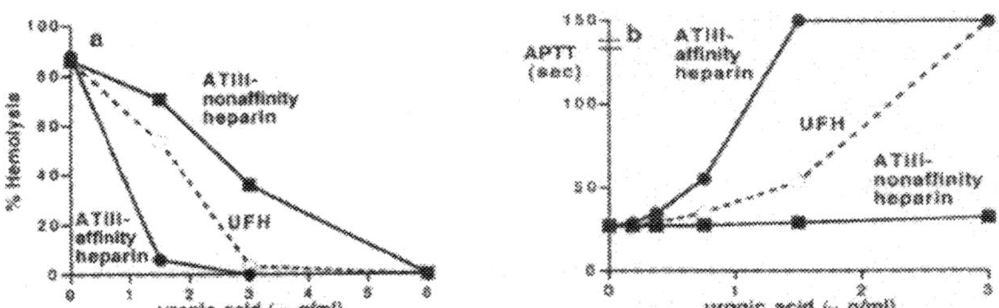

Fig. 2 Independency of anti-C' and anticoagulant activity of heparin. ATIII-affinity and –nonaffinity heparin was separated by ATIII-affinity column. **a** ATIII-nonaffinity heparin showed a significant anti-C' activity, although less than that of unfractionated heparin (UFH) or ATIII-affinity heparin. **b** ATIII-nonaffinity heparin showed little anticoagulant activity.

Anti-C' and anticoagulant activity of heparin (Fig. 1)

C'-mediated lysis of PNH RBC enhances the coagulation cascade of the plasma. Heparin or LMWH effectively inhibits the both pathways of C' and coagulation, however, IC_{50}s of heparin as anticoagulant was about 10-fold less than that as anti-C'.

Anti-C' activity of anticoagulantly-inactive heparin (Fig. 2)

ATIII-nonaffinity heparin fractions showed significant anti-C' activity. N-desulfated, acetylated heparin (NAcH), anticoagulantly inactive, also showed significant anti-C' activity (data not shown).

Discussion

Whereas UFH or LMWH can inhibit C'-mediated lysis of PNH RBC in vitro, the concentrations required to inhibit C' efficiently were much higher than those obtained clinically [6]. The 10-fold difference in the activity of heparin between as anti-C' and as anticoagulant (**Fig. 1**) also indicates the commercially-avaliable heparin preparations could not directly be applied clinically by expecting their dual effects. Indeed, 2 courses of heparin- or dalteparin-administration at clinical dosages did not obtain in vivo concentrations sufficient to inhibit C'. AT III-nonaffinity fractions of heparin or its derivatives with significant anti-C' activity are potentially useful to inhibit C'-mediated lysis of PNH RBC, although adverse effects other than the anticoagulant effects should be studied carefully before clinical usage.

References

1. Jacques LB (1979) Heparin: an old drug with a new paradigm. Science 206:528-533
2. Lam LH, Silbert JE, Rosenberg RD (1976) The separation of active and inactive forms of heparin. Biochem Biophys Res Comm 69:570-577
3. Caughman GB, Boackle RJ, Vesely J (1982) A postulated mechanism for heparin's potentiation of C1 inhibitor function. Mol Immunol 19:287-295.
4. Weiler JM, Edens RE, Linhardt RJ, et al (1992) Heparin and modified heparin inhibit complement activation in vivo. J Immunol 148:3210-3215.
5. Rosse WF, Ware RE (1995) The molecular basis of paroxysmal nocturnal hemoglobinuria. Blood 86:3277-3286.
6. Ninomiya H, Kawashima Y, Nagasawa T (2000) Inhibition of complement-mediated haemolysis in paroxysmal nocturnal haemoglobinuria by heparin or low-molecular weight heparin. Brit J Haematol 109:875-881.
7. Bitter T, Muir HM (1962) A modified uronic acid carbazole reaction. Anal Biochem 4:330-334.

A Patient with Paroxysmal Nocturnal Hemoglobinuria in whom G-CSF Administration was Remarkably Effective against Recurrent Gastrointestinal Infections and Hemolytic Episodes Caused by Cellular Immunodeficiency

T. Matsunaga , A. Fujimi , N. Takemoto , I. Tanaka, T. Akiyama, S .Sakamaki and Y. Niitsu

4th Department of Internal Medicine, Sapporo Medical University School of Medicine, South-1, West-16, Chuo-ku, Sapporo 060-0061, Japan

Summary: We report an elderly patient with paroxysmal nocturnal hemoglobinuria (PNH), having recurrent gastrointestinal infections and hemolytic episodes caused by cellular immunodeficiency due to a decreased number of T cells. On admission to our hospital, the patient had normal neutrophil count and function, but had decreased T cell count, decreased mitogenic reaction to PHA and ConA, and negative reaction to the tuberculin test. Thus, a cellular immunodeficiency state was suggested and postulated to be due to decreased T cell count caused by the loss of PNH clone of T cells. We previously reported that the administration of G-CSF to normal donors of allogeneic peripheral stem cell transplantation increases the lymphocyte count. Based on this experience, we administered G-CSF to this patient. This resulted in an increase in the T cell count, normalization of the mitogenic response to PHA and ConA, positive reaction to the tuberculin test and increases in the blood levels of Th1 cytokines and Th2 cytokines, with an improvement in the gastrointestinal infections and hemolytic episodes. This suggested that G-CSF administration resulted in the recovery of cell-mediated immunity in the gastrointestinal mucosa. At 9 months after the start of G-CSF administration, these complications have not recurred. This is the first report of a patient of this type and suggests that G-CSF administration may be useful in the treatment of elderly PNH patients with cellular immunodeficiency.

Key words: Paroxysmal Nocturnal Hemoglobinuria, G-CSF, Cellular immunodeficiency

Case report

A 67 year-old woman was presented on Feb 10, 2000 with a fever of 38°C, abdominal pain, and malaise at Naganuma Municipal Hospital. The patient was diagnosed with acute gastroenteritis and treated with antibiotics and antifungal agents. The symptoms improved, but discontinuation of treatment resulted in a recurrence of the same symptoms. Furthermore, because of coexistent hemolytic anemia, the patient was referred to our hospital on June 14, 2000, for further evaluation and treatment. At the time of admission, the palpebral conjunctiva showed a mild degree of pallor, the fever was 37.8°C, and there was slight tenderness to palpation in the lower abdomen. The results of the examination at the time of admission to this department were as follows. The CBC showed leukocytosis due primarily to neutrophils, normocytic normochromic anemia, thrombocytopenia, and reticulocytosis. Coagulofibrinolysis studies showed elevated FDP, D-dimer and TAT, suggesting a hypercoagulable state. Blood chemistry studies showed LDH elevation, while serum studies showed decreased haptoglobin and positive anti-candida antibody. Urinalysis was positive for blood and positive for urinary hemosiderin, and the stool was also positive for occult blood. Based on the physical and laboratory findings, the patient was suspected of having gastrointestinal infections and anemia due to intravascular hemolysis. In the evaluation of the intravascular hemolysis, the Ham and sugar water tests were positive, establishing the diagnosis of PNH. The expression of GPI-anchored proteins was determined by measuring the CD55 and CD59 by two-color flow cytometry on the red cells, lymphocytes, monocytes and polymorphonuclear leukocytes in the peripheral blood. All blood cells showed a CD55-negative cell fraction and CD59-negative cell fraction. These findings confirmed the diagnosis of PNH.

The clinical course is described in Figure 1. The patient had been admitted in February 2000 and antibiotics and antifungals were given for the diagnosis of acute gastroenteritis by the preceding physician. The symptoms improved, but discontinuation of treatment led to the recurrence of similar symptoms. When admitted to our department in June 2000, the patient defervesced with continued administration of antibiotics and antifungal agents. However, irritation of the gastrointestinal tract from the gastrointestinal tract fiber scope (GTF) and colon fiber scope (CF) performed for the purpose of gastrointestinal evaluation resulted in the recurrence of fevers and hemolytic episodes, despite continuation of the above treatments. During the febrile episodes, the neutrophils count increased, but the lymphocyte count remained below 1,000/μL. Repeated transfusions were required for the hemolytic episodes. GTF showed candidal duodenitis, while CF showed non-specific colitis. Microbial cause of the colitis could not be identified. Immunological data for this patient were as follows. At the time of admission the patient showed decreases in peripheral blood total lymphocyte count, CD3 positive cell count, CD4 positive cell count, CD8 positive cell count, and CD19 cell count. Therefore, T cells and B cells decreased. The decreased lymphocyte response in the mitogenic test with PHA and ConA and negative reaction to tuberculin testing suggested decreased cellular immunity. These data were consistent with the decreased T cell count.. Helper T cell (Th cell) subsets were enumerated by the method of Kanegane et al. [1]. CD45RO and CD62L were analyzed by two-color flow cytometry. It was observed that the peripheral blood Th1 and Th2 cells both decreased compared to the levels in healthy adults. Furthermore, Th cell cytokine production was measured. ConA was added to 1 x 10^6/mL of peripheral mononuclear cells, cultured for 24 hr, and the supernatant was analyzed by the ELISA method. When compared to healthy adults, the IL2 and IFN-γ, which are Th1 cytokines, were similar, but the IL-6 and IL-10 Th2 cytokines were low in this patient. These data suggested that in this patient, the loss of the PNH clones of T cells, resulting in decreased T cells, led to a cellular immunodeficiency state and repeated opportunistic infections of the intestinal tract. On the other hand, the white cell phagocytic function, a parameter of neutrophil function, was normal.

We have previously reported that the administration of G-CSF to healthy adult donors of allogeneic peripheral stem cell transplantation resulted in increases in the neutrophils, as well as lymphocytes [2]. In order to increase the T cells in this patient G-CSF was started on October 6, 2000 to see if cellular immunodeficiency could be improved. Specifically, G-CSF was administered twice a week at 150 μg/dose to maintain the neutrophil count at 10,000/μL. It was observed that the patient showed an increase in the neutrophil count, resolution of the fever, and hemolysis, and became independent of red cell transfusions. More surprisingly, there were increases in the T cell count, as well as the B cell count. GTF performed on Nov 30, and CF performed on December 10 showed resolutions of the abnormal findings. At 3 weeks after the start of G-CSF administration, the antibiotics and antifungal agents were stopped. After 9 months of observation under therapy with G-CSF alone, the gastrointestinal infections and hemolysis have not recurred. Hematologic parameters were measured after G-CSF administration and compared to pre-G-CSF administration data. As already indicated in Figure 1, the peripheral blood lymphocytes increased after G-CSF administration. Further analysis of the lymphocytes indicated increases in both the T cells and B cells. In addition, the tuberculin test, which showed negative at the time of admission, became positive after G-CSF administration. Therefore, the G-CSF administration corrected the cellular immunodeficiency in this patient. When compared to the state prior to G-CSF administration, the peripheral blood Th cells, both the Th1 cells and Th2 cells, increased. The cytokine production by peripheral blood Th cells, when compared to pre G-CSF administration, also increased for both Th1 cytokines and Th2 cytokines. The Th2 cytokines IL-6 and IL-10 were especially greatly increased compared to the Th1 cytokines. The PNH clones of red cells, lymphocytes, and polymorphonuclear leukocytes with no CD55 or CD59 expression, as well as the normal clones, increased after G-CSF administration. Two possible mechanisms for the increase in the red cell count and the lymphocyte count are G-CSF mediated stimulation of the cells at the levelof hematopoietic stem cell and the improvement in the loss of PNH clones resulting from improvement in the infections. The G-CSF administration lead to an increased number of nucleated cells and an elevation in the M: E ratio in the bone marrow.

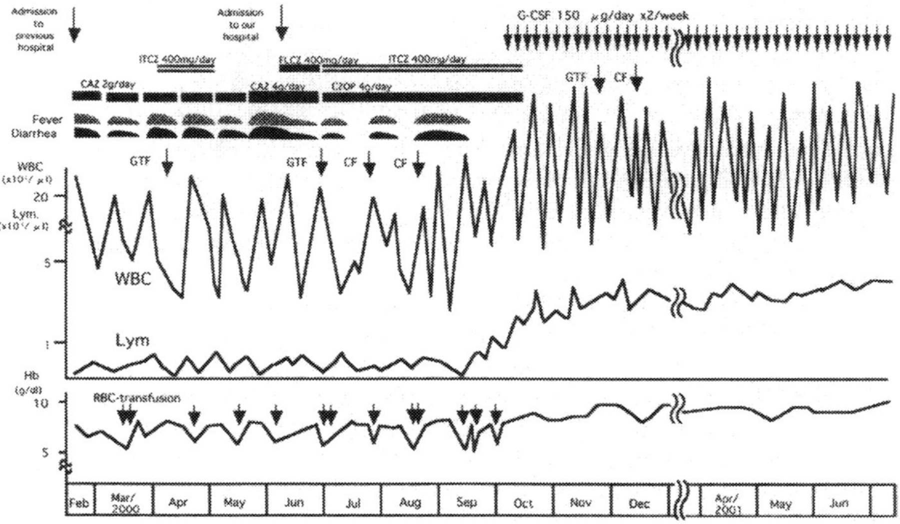

Fig.1. Clinical course. ITCZ, Itraconazole; FLCZ, Fluconazole; CAZ, Ceftazidime; CZOP, Cefozopran hydrochloride; G-CSF, granulocyte-colony stimulating factor; GTF, Gastrointestinal tract fiber scope; CF, Colon fiber scope; WBC, White blood cell; Lym, lymphocyte; RBC, Red blood cell; Hb, Hemoglobin.

References

1. Kanegane H, Kasahara Y, Niida Y et al (1996) Expression of L-selectin(CD62L) discriminates Th1- and Th2-like cytokine-producing memory CD4+ T cells. Immunology 87: 187-190
2. Matsunaga T, Sakamaki S, Kohgo Y et al (1993) Recombinant human granulocyte colony-stimulating factor can mobilize sufficient amounts of peripheral blood stem cells in healthy volunteers for allogeneic transplantation. Bone Marrow Transplant 11: 103-108

Key Word Index

MIX
Papier aus verantwortungsvollen Quellen
Paper from responsible sources
FSC® C105338

If you have any concerns about our products,
you can contact us on
ProductSafety@springernature.com

In case Publisher is established outside the EU,
the EU authorized representative is:
Springer Nature Customer Service Center GmbH
Europaplatz 3, 69115 Heidelberg, Germany

Printed by Libri Plureos GmbH
in Hamburg, Germany